Seeing Animals

Seeing Animals

Their Path Through Our History

Angela Dyer

Ⓛ
The Lutterworth Press

The Lutterworth Press
P.O. Box 60
Cambridge
CB1 2NT
United Kingdom

www.lutterworth.com
publishing@lutterworth.com

Paperback ISBN: 978 0 7188 9542 6
PDF ISBN: 978 0 7188 4778 4
ePub ISBN: 978 0 7188 4779 1
Kindle ISBN: 978 0 7188 4780 7

British Library Cataloguing in Publication Data
A record is available from the British Library

First published by The Lutterworth Press, 2019

I think I could turn and live with animals . . .

Walt Whitman, 'Song of Myself' 32

Contents

Illustrations

Introduction

Seeing animals as they are, without ulterior motives of possession or power or gain – economic, scientific or even artistic – is liberating to the one seeing, possibly also to the seen. To see something, it is not enough merely to look. Seeing allows us also to be seen, opens us up to the possibilities of what we look at, and how we might respond. If we *see* with our whole being, we are likely to find the essence of whatever creature we come upon. Binoculars or microscopes may not help, but empathy does: the heart must be involved as well as the mind.

Our ancestors *saw* the animals they drew on their cave walls, and though they hunted them, they also revered and then came to worship them. Animals were gods, messengers from another world – and look what we are doing to the gods now. Maybe we should look again, and learn to see.

Because we too are animal, the way in which we treat animals reflects how we treat ourselves. If we lock them up, exploit them, treat them with cruelty or indifference, we are doing this to ourselves and to each other. Containment takes many forms, and one of the most damaging aspects of lack of freedom is loss of desire. A contained animal, whether it is behind bars in a zoo or shut up in a flat all day, faces no danger, has no choices to make, can do nothing but repeat the same movements, over and over, round and round. Having lost the ability to explore and to face the unexpected, it will eventually – because animals are realists – become dulled and dispirited, turning in on itself. It becomes, like many of its human counterparts in a regimented urban environment, depressed.

Being outside with a dog that is free, in a field, on a hillside, in woodland, observing it as it surveys its surroundings, ears pricked, eyes searching, nose twitching at the myriad signals it picks up while deciding what to do next, where to wander or dash – this is a sight that makes us too feel alive, helps us to see the world with an intensity that is largely absent from the daily round. It awakens our animal nature and produces a sense of wellbeing, a hint that perhaps all's well with the world after all – after all we humans have done to wreck it. Looking at a dog on a television screen, or even at your own curled up on the hearthrug, does not have this effect. In order to empathise with a dog (or any creature), you need to be out there with it as it sniffs and stalks and suns itself, doing what it was born to do when in control of its own life, however fraught and short that may have been.

Seeing is very different from watching. To *see* one must accept whatever one is watching, whether animal, fellow human being or anything else, in its own reality rather than through the filter of one's preconceptions or needs. No one could quarrel with bird-watching, which besides being a tranquil hobby provides useful information on numbers and rarities, but it may add up to little more than trainspotting, notation rather than involvement. Can one be involved with a bird in the wild? Perhaps not, but one can see it as bird, not breed, an individual embarking on its perilous voyage across continents, risking the hazards of exhaustion, lack of food, bad weather and predators, its odds against survival so heavily stacked. Or with numbers of birds? You can watch a murmuration of starlings, and wonder at the why and how, but if you are not moved– by its calligraphy, its choreography, whatever image comes to mind – something is missing.

Identifying with animals does not, emphatically, mean treating them in a sentimental, indulgent way. Killing with kindness is arguably more offensive to an animal's integrity than killing it as a sacrifice or simply to eat. The anthropomorphism that led ancient man to see animals as equals added to their dignity; dressing them in tutus does not. Animals have a

Starlings make moving pictures in the sky, but the stills take on different forms.

natural sense of dignity, which is not to say that they have no sense of fun, but that they will try at all costs to preserve their self-respect, and we should in turn respect this. Making fun of animals, as distinct from having fun with them, is a betrayal of their trust in us.

Learning what we as humans have done both to earn and destroy that trust throughout our short history together is a useful mental exercise, but it has to be transformed into a leap of the heart if we are to appreciate the essence of animals, and why they matter to us.

* * *

The exchange of greetings between the San people of the Kalahari translates as 'I see you'. It's a lesson we could learn, since in the rush of modern life it is very easy *not* to see, and not seeing becomes a habit – in fact one is advised to avoid eye contact in order to keep out of trouble in cities. But having myself abandoned the press of city life for a rural backwater in Brittany, I still at times do not see, hurrying to let out the chickens in the morning without pausing to notice the bantam cockerel: his dazzling flame-coloured feathers cascading from the neck and merging with the russet ones that fall over his back, the tiny powder-puff of white between this and the high spout of his tail, at first glance

black but on closer observation deepest blue-green. And these are only
the externals. What about acknowledging his status as cockerel, his need
to climb higher than all the others and crow, yet his solicitousness for
the flock as he calls them to share a cache of juicy worms; he's a cocky
gent, this fine bantam. The close observation often doesn't take place, his
beauty and his characteristics are taken for granted – and I am the loser.

Seeing, rather than merely looking, means focusing on the object not
as object but as fellow subject, as individual: this cockerel, this cockroach,
this dog or dogfish, by chance alive in the world with me at this moment,
each of us personifying and perpetuating the life of our own unique
species. It may not, probably won't be a communication, but something
takes place that immeasurably enriches the seer, a word that once meant
receiver of divine inspiration.

* * *

What of the animals, what do they see? Science can tell us about the
physical structure of eyes, about numbers and ratios of rods and cones,
peripheral sweep and night vision, and we can therefore surmise many
differing abilities over a range of species, from owl to mole to bat (not
blind at all). But the mechanics of sight is not our subject; to look is not
necessarily to see. Animals use their eyes in tandem with smell and hearing,
all three closely linked to intuition – and all with the urgent, primary
focus of avoiding danger and finding food. These senses are as superior to
those of humans as our language is superior to theirs. Uncluttered by the
judgements and controls imposed by our rational 'modern' brain, animals
react instinctively, picking up on the emotional energy of other beings,
animal and human, and instantly aware of any threat. Such intuitive
sensing has become blunted in so-called civilised adult humans though it
remains stronger in young children and people who live close to Nature.
We get a glimpse of this on the rare occasions when we 'sense' danger –
though it may not be visible – and before the mind has even registered it,
the adrenaline pumps in, the pulse quickens, the mouth goes dry and we
are on alert, a survival instinct that links us to our animal forebears.

As to what animals see when they look at us, we can only imagine. Are
we no more than blurred outlines, objects of fear or affection, providers of
food and shelter? That many animals recognise humans, even after long
periods apart, is irrefutable. Donkeys have been known to greet former

owners after more than twenty years, and the video of the lion Christian, bought in Harrods and released into the wild in Africa, falling into the arms of his rescuers several years later, is an all-time YouTube favourite. How much is visual, how much response to voice, how much sensing or just memory click we cannot know, although animal behaviourists are busy working on it. In most cases, the animal in question has known the person from infancy and has imprinted on him or her. And what of that imprint? This is surely a most intense sort of seeing, an indelible modelling on the parent, whether the biological parent or a surrogate, that lasts a lifetime.

It's humbling to imagine how we appear to animals. A while ago I stopped on the edge of a small town in Brittany where a travelling circus was parked; how bizarre, how delightful to come across a couple of llamas, a camel and several miniature ponies, tethered and grazing outside a supermarket. But the delight turned to dust when I came to the lions, cooped inside a metal trailer, listless and immeasurably sad. I cannot forget the way they looked at me, looked *through* me, with dignified disdain, as if I were beyond fear or hatred, merely of no interest – this torpor the deepest wound of incarceration. It may be misleading to anthropomorphise, to imagine how we would feel if we were an animal bred for the savannah spending our life in a trailer, yet how else can we be aware of what we do to animals or wonder why we do it?

Another big cat, the elusive and solitary leopard that occasionally turns to human prey – the 'man-eater' of many a tale – appears to *see* humans. Peter Matthiessen relates that 'a leopard will lie silent even when struck by stones hurled at its hiding place – an act that would bring on a charge from any lion – but should its burning gaze be met, and it realizes that it has been seen, it will charge at once'. Many animals are reluctant to meet the gaze of humans, and even dogs may become uncomfortable and look aside. Dogs, and surprisingly horses too, are able to distinguish between a loving and a hostile look, and this account of a visual conversation between J.R. Ackerley and his dog Tulip shows how powerful the communication can be, though here it is the man who is discomfited: 'The tall ears are erect now, the head drawn back, the gaze level. I meet it, in spite of myself. We stare into each other's eyes. The look in hers disconcerts me, it contains too much, more than a beast may give, something too clear and too near, too entire, too dignified and direct, a steadier look than my own. I avert my face.'

We don't know for sure whether animals pine for a lost Eden, but many people have addressed the possibility. An abandoned wolf cub, reared as a pet, becomes uneasy and begins to stare into the distance, then one day it vanishes; the Chippewa Indians described this as 'the sickness of long thinking', the yearning to return to the wild. Laurens van der Post similarly writes that horses have 'a sadness glowing at the far end of the long look' and are 'haunted by dreams of their birthright of freedom exchanged for a mess of oats and the security of luxurious stables'.

The eyes of the equines often have what we describe as sadness, for just as our eyes are a window to the soul, so can animals' eyes be expressive in ways that we may interpret through human emotions. As D.H. Lawrence put it, 'Elephants in the circus have aeons of weariness round their eyes.' 'Eyes, always eyes' writes John Berger, as he observes the apes in a zoo in Basel; and it was the eyes of London Zoo's perhaps most famous inhabitant, the gorilla Guy, that captivated his audience – eyes that were sad, yes, but intense and lit with intelligence, eyes that seemed to communicate with humans from a deep well of shared wisdom.

'Eyes, always eyes'. The status of this western lowland gorilla is as emphatic as its presence: its scientific name is Gorilla gorilla gorilla.

* * *

In general, people today have become accustomed to seeing animals not as they are. In books, films, cartoons and advertisements, quadrupeds in human clothes walk on two legs and are attributed with human speech and emotions; they are people substitutes, often caricatures. This may appear harmless, just another of the 'legitimate' uses we in our arrogance find for animals. But at best it is tinged with ridicule, and at times comes perilously close to abuse, of dignity and of character.

We have got used to seeing animals on screens and behind bars, their jaws and claws safely contained, their nature and needs denied. Dogs are on leads; horses harnessed; cows live on production lines; and chickens whose pumped-up Page 3 breasts are now brazenly displayed on plastic trays in the supermarket have never known what it was to scratch the earth, eat worms or establish their rightful pecking order. Most of these animals are no longer visible to the vast majority of humankind, hurrying about its business on tarmac and concrete.

The aim of this book is to encourage the reader to slow down and notice animals, go out and look for them and look again, in order really to see them – not as mere objects for our use, nor as subjects of study, but as individual creatures who share with us the mystery of life on Earth. Being conscious of the part animals have played in the history of human beings, from very earliest times up to the present, may open one's eyes to what is easily overlooked in our largely urban lifestyle, namely their importance to humans on so many different levels, from the strictly practical – and what can be more practical than providing us with nourishment – to the mythical and artistic. For if we don't *see* animals, we are unlikely to give them the respect they deserve, and if we can't find a way to live at peace with them, how can we be at peace with ourselves?

I
WORSHIPPING
ANIMALS

The word 'animal' means possessing vital breath or spirit, a short step from the spirit world. The religious significance of animals has been a constant throughout history; in the unceasing quest of human beings to make sense of their world through stories and images, what more vivid material could there be for the imagination than the multiplicity of forms, colours and behaviour that make up the animal kingdom.

Animals carry our needs and desires, and it is precisely because they cannot speak our language that they represent something pure, uncontaminated by human duplicities and conceits. Take birds, those airy spirits soaring heavenwards, defying gravity and leaving behind the land to which we are bound. It's easy to see why they were perceived as messengers from the gods, or the dead, and why they are the stuff of myth. Who hasn't longed to climb on the eagle's back, to cast off from the cliffside or float upwards on a thermal? While watching birds and *seeing* them, one becomes for the moment a free spirit with them.

Identifying with animals is as close as we can come to the early relationship between animal and human that is hypothesised. Of the Kalahari Bushmen, or San people, Laurens van der Post writes that they and the animals 'participated so deeply of one another's being that the experience could almost be called mystical. For instance, [they] seemed to *know* what it actually felt to be an elephant, a lion, an antelope. . . . it seemed to me that [the San's] world was one without secrets between one form of being and another.' This is not an airy-fairy Garden of Eden, but one in which humans survived by killing and eating animals. For it seems likely that primitive men identified so closely with animals for this very reason: they had to know them in order to be successful as hunters. Indeed some believed that in eating the meat of an animal, they incorporated its being into their own; they became that animal.

Thus animals must be hunted and killed, but in this they must also be venerated. Veneration led to totemism and on to deification, the mystery of animals allied to the mystery of gods: that which cannot be understood must be propitiated and worshipped. In many cases the animal was not worshipped – though it may have been revered – for its own sake but because it was thought to represent, or even to manifest, a god. Hindus, for example, who refrain from eating cows and even to this day allow them to wander their traffic-bound streets, do so from a long-held tradition of respect for the 'sacred cow' as goddess and mother rather than for love of the beasts themselves.

* * *

Worship, reverence, respect – all words that refer to a space between the subject and object. There is no place for over-familiarity here, certainly none for indulgence; these are lean words describing a serious state of mind on the part of the worshipper. As to the animals, they remain at all times dignified, a little aloof, worthy of worship.

So let's take a look at some of the many animals that have been worshipped by humans over the millennia. Those San of southern Africa made studies on rock, both paintings and engravings, of a power and technical mastery that equals or even excels the more widely known European cave paintings. Certainly the variety of animals depicted is greater, and the San could not have painted these animals – eland and other antelope, cattle, horses, elephants, rhinos, hippos and many other mammals, as well as birds and bees, snakes and the occasional crab – so accurately without knowing how they were put together, what lay under their skin. And, as Leonardo da Vinci was to show thousands of years later, there is no better way of finding this out than by dissecting, or in the case of the San, butchering them. But these African animals were not mere meat and were, it now appears, more than just objects of wonder. Recent study of African rock art shows that these images, packed as they are with symbolism and metaphor, reflect without any doubt the role of animals in religious beliefs and practices and the link they were believed to provide between the physical and spirit worlds.

Central to the belief of the San was that the altered state of consciousness achieved by the shamans in their 'trance dance' put them in touch with the spirit world and allowed them to cure disease, foretell the future and perform other supernatural acts. And central to this, as messengers and go-betweens, were animals. A shaman often identified with a particular animal, which would act as his spirit guide and provider of 'potency'. The eland was the most highly prized and most invoked of the ritual animals, as it was believed to be especially endowed with this potency. The shaman possessing eland power would be able to see what eland saw, feel what they felt and know where they were at any given time; useful knowledge also for the hunter. This connection was confirmed by a San less than fifty years ago, who described 'hunting techniques and rituals as if there were no difference between the two'. The healing dances were often combined with the killing and sharing of an animal for food in a joint celebration: van der Post was told that 'ever since the days of the first Bushman, no hunter had ever killed an eland without thanking it with a dance'.

The eland gains its status not just on account of its size and beauty, but because of its fat. To all hunter-gatherers fat is highly desirable, but eland fat is not only prized as food. The supernatural potency tapped into by the shaman is believed to reside specifically in the fat, which is also used in initiations, rites of passage and menstruation ceremonies. In the trance dance the shaman trembles, sweats profusely, staggers and bleeds from the nose as he approaches entry to the spirit world, actions that mimic the behaviour of an eland as it dies.

Besides the four-footed animals that one can almost smell as they pound across the painted rocks of southern Africa, there are some spectacular birds – and some engagingly comical ones. For the San, as for us all, flight is a metaphor for escaping our earthbound limitations, and nothing evokes this so well as the soaring flock of birds at Uitenhage in the Eastern Cape, their ochre forms silhouetted against the cloud-like grey rock. The trance experience was identified with the sensation

Flying, soaring, wheeling – the birdlike sensations associated in San mythology with spirituality are personified in a rock painting from The Drakensberg, South Africa.

of weightlessness, which is likewise vividly captured in the rock painting illustrated. But it was the very earthbound ostrich that fulfilled so many needs for early Africans, providing them with eggs to eat and the shells as containers, currency and jewellery; with meat and oil; with feathers and leather. These too were surely worshipped, but we have no evidence of this other than a wry caricature on a rock face.

* * *

Birds had a role in the religion of the ancient Egyptians. Horus, god of the sky, was portrayed as a man with the head of a hawk or sometimes simply as a hawk. He is best known for his single eye, lost in battle but later restored, and adopted by the ancient Egyptians as a symbol of protection. The Egyptians didn't worship birds or animals per se, but believed that the gods resided in animals and accordingly treated them with great reverence, often mummifying and burying them alongside their pharaohs: pets to keep their masters company in the afterlife, along with other animals to provide them with food.

Another ancient Egyptian god with a man's body and a bird's head was Thoth, usually depicted with the head of an ibis, though sometimes, confusingly, with that of a baboon. Thoth, god of equilibrium, acted as a mediator in disputes, as well as being credited as the inventor of writing and a dazzling array of sciences including mathematics and astronomy, 'the true author of every work of every branch of knowledge, human and divine'. One wonders how the little ibis head could hold so much and how its attribution came about; the Egyptians regarded the long curved beak as symbolic of the crescent moon, which was associated with wisdom, but more prosaically such a beak must surely have been useful for picking up all those pieces of knowledge.

Thoth, here with the head of an ibis, on a relief in the temple of Ramesses II in Abydos, Egypt.

The giant of the Egyptian pantheon was the sphinx, possibly better equipped for intellectual capability with its human head, set upon the flanks of a lion and sometimes also bearing wings. But though in Egyptian mythology sphinxes were massive, benevolent pussycats acting as guardians to the tombs of the pharaohs, the feminine Greek version is altogether more threatening: a monster with female head, lioness's body, eagle wings and serpent's tail, prone to setting obscure riddles and devouring all those who fail to answer them. This Sphinx had a predilection for young men, but finally met her match in Oedipus and flung herself off a cliff.

The Greek gods had an intimate relationship with animals, and none more so than Zeus. His sacred animals were the bull and the eagle and he used both to his advantage, seducing Europa (who subsequently gave birth to Minos) in the guise of a bull, and as an eagle abducting the beautiful youth Ganymede. Zeus took advantage of his sister Hera's soft spot for animals by disguising himself as an abandoned baby cuckoo before jumping on her, but it was as a swan that he performed his most dramatic seduction – or was it rape? In perhaps the most famous rendering of the myth of Leda and the swan, there is little doubt. Yeats's poem needs no illustrating; it is one of the most vivid and explicit descriptions in poetry:

How can those terrified vague fingers push
The feathered glory from her loosening thighs?

Leda succumbs, and thus is conceived Helen of Troy.

* * *

Animals don't necessarily have to embody or even represent gods in order to play a part in spiritual belief. Animism is a broad term for the belief system shared by indigenous people before the advent of formal, god-based religion; less a religion than a way of viewing the world, making sense of it, as people explored what it was to be alive. For them there was no division between the spiritual and the physical dimension, and souls were shared by humans and animals alike, an inclusive, 'innocent' attitude towards life closer to the idea of paradise before the arrival of the serpent.

In *The Golden Bough* Frazer wrote of the people of Calabar, in southern Nigeria, who 'believe that every person has four souls, one of which always lives outside of his or her body in the form of a wild beast in the forest. This external soul, or bush soul . . . may be almost any animal, for example, a leopard, a fish, or a tortoise; but it is never a domestic animal and never a plant.' He goes on to say that the life of the man is so intimately bound up with the animal, which he regards as his bush soul, that death or injury of the animal may also bring about death or injury of the man. 'And, conversely, when the man dies, his bush soul can no longer find a place of rest, but goes mad and rushes into the fire or charges people and is knocked on the head, and that is an end of it.'

The beliefs of the Australian Aborigines are centred in the land, and they regard animals and plants as the creators of that land. In their Creation stories, the world was flat and barren until the Rainbow Serpent awoke from its sleep and pushed through the surface of the earth, creating lakes and rivers and giving life to animals, plants and humans, who then helped to form the land. This gives to all a common kinship and a common soul; reincarnation allows all living forms to intermingle and exchange souls. From this belief comes an inevitable respect for all fellow beings: you do not abuse your family, and especially not your ancestors. It's a complex set of beliefs not to be explored here, but the inclusion of animals (and plants) in the hierarchy and their status as equals with humans has an obvious effect on how they are regarded and treated.

In the formative time of the Creation, which the Aborigines called the Dreamtime, the various species were believed not yet to be fully formed, the boundaries between human and animal remaining undefined. In what came to be known as totemism, a person would take on a special relationship with one particular species or individual animal, his totem. Some totems are not chosen but bestowed at birth or even conception, and this totemic being, most often an animal, guides and influences the child, often appearing in dreams.

* * *

The Aborigines of Australia are not alone in having totem animals. For most people the totem image that comes immediately to mind is the carved pole of the North American Indians – huge, colourful and teeming with stylised birds, animals and fabulous beasts representing the family of the chieftain – but there are more subtle examples that also delineate clan. The Tlingit, an indigenous people of the Northwest Pacific Coast, were animists and carved intricately designed bowls, utensils and ceremonial rattles incorporating ravens, oystercatchers, frogs, bears and other totem animals. They also portrayed deliberately ambiguous creatures spanning the 'moieties', or divisions, of Tlingit kinship that blurred the boundaries between one animal species and another. Describing a bowl carved from the horn of a mountain sheep and decorated with abalone shell, portraying an animal that could be a wolf but equally a marmot, the anthropologist Robert Storrie writes: 'Trying to decide exactly what creature is carved on the bowl is probably to misunderstand the carver's intention. At an important level the design can

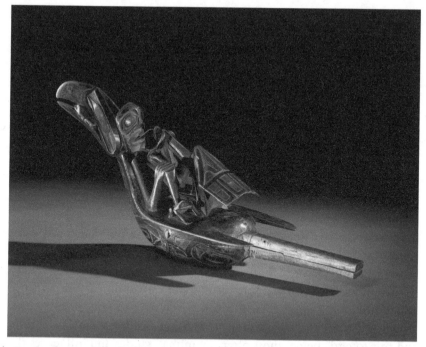

Rattles depicting ravens and other animals, here frogs, are used in ceremonial dances to demonstrate the power of chiefs throughout the Northwest Coast. A shaman reclines on the raven's back, and the sound of the rattle is believed to form a conduit to the supernatural world.

represent the transformational potential of all beings. Ambiguity acts to give the bowl a blank character in which changing identities can be discovered. They can be wolves for me and marmots for someone else, yet neither of us is wrong.' A good lesson for us in the twenty-first century, with our obsessive need to classify and pigeon-hole.

Relating to a particular animal and identifying with it, even emulating it, is common to many societies and is still practised today. Of the twelve signs of the astrological zodiac familiar to Westerners, seven (and a half if you count Sagittarius) are animals. And most of us will have found similarities – in our friends, if not ourselves – to the animal characteristics of the star signs: a stubborn Taurus with his feet solidly on the ground, that slippery Pisces you can't get hold of, an Aries trampling everything in sight. It's surprising how often the traits fit. The Chinese astrological signs are all animals, the twelve that came to bid farewell to Buddha before he died, after each of which he named a year. And anyone interested in finding their own 'spirit animal' can have a ball on the internet – though you might do better going off into the woods and sitting under a tree for a few hours.

* * *

The Buddha is reputed to have sat under a tree for forty-nine days in order to achieve enlightenment, and it is tempting to speculate on the insects and animals that would have visited him during his still and silent vigil; it seems likely that in his heightened state of consciousness he would have *seen* them, been truly aware of their essence. And there would have been many, for the tree he sat under was a pipal tree, a sacred fig, and figs are dependent on wasps in a complex system of mutual gain, while the fruit itself feeds an array of exotic birds such as orioles and barbets as well as a variety of mammals. What symbolism here for anyone with the time, and the eyes, to see.

The Buddhist creed insists on the sanctity of life for all sentient beings, with no division between human and animal. This means that taking life for food is not allowed and the eating of meat is forbidden; animals must be treated with compassion and empathy at all times. So far, so good. Yet Buddhists also believe that humans are reincarnated as animals because of misdeeds in this life, which immediately implies the inferiority of the animal even if it is held that both human and animal have the potential for enlightenment, the ultimate goal in Buddhism. Leaving aside this somewhat obscure ideology, the avoidance of violence, however theoretically desirable, appears far from reality when one considers the law of 'kill in order to survive' that maintains most of the animal kingdom. But if it reduces by any tiny percentage the suffering inflicted by our current methods of meat production, then it is surely to be encouraged. If Buddhism has much to say about the sanctity of animal life, it offers little about individual animals. There is though a touching tale about the Buddha himself: in a previous lifetime he was said to have sacrificed his own life in order to feed a tiger and her two cubs, trapped by the snow and starving, his reasoning being that it was better to save three lives than to preserve his own.

Hindus share many of the attitudes of Buddhism regarding the sanctity of animal life. Spiritually, Hinduism sees no distinction between animals and humans: all possess souls and all are manifestations of God, though with animals as limited beings and on a lower scale of evolution than mankind. Many animals are worshipped, particularly cows that are considered sacred and must never be harmed, though they may be killed as sacrifice to the gods. This may seem at best perverse, but the reasoning is that as we depend on the gods for our protection, so the cows depend upon us for their welfare, and as we nourish the gods with sacrifice, so cows nourish us with their milk. It's a strange logic to Western ears, but one that we who keep cows in a metal stand for their entire lives should ponder before we judge it. The practice of sacrifice is denied by many Hindus and is now dying out as attitudes change. But in India, cattle past their prime as milk producers have always been revered and allowed to die naturally and now, rather than competing with the increasing traffic on city streets, they are more likely to be taken off to the bovine old-age homes that are springing up across the country. In fact, a recent report states that 'funds for old cows may outstrip that [*sic*] for senior citizens'.

Cattle have been, and still are, the literal lifeblood of many African people; their dependence on cattle is absolute. The Dinka of East Africa are particularly known for devotion to their cows, as their lives are inextricably bound. They don't eat them as meat until the cows die naturally, but live almost exclusively on their blood and milk. But this is only a fraction of what cows provide. Their dung is used as fuel and plaster; their hides for leather; their bones for making utensils; their horns for spoons, spears and ornaments; their scrotums for pouches; the hair from their tails for tassels. These are merely the physical gains. Cattle are also exchanged as currency and dowry, battles are fought over them, social gatherings centred upon them, spiritual beliefs vested in them. There's a story about a Christian missionary who, extolling the joys of heaven to a group of Dinka, is asked by an elder if he can take his cattle with him when he goes to heaven. The preacher fluffs the answer, so the elder shrugs and says that without his cattle, heaven isn't worth going to. You have to remember that these cattle of the southern Sudan are elegant beasts far removed from our chunky farmyard cows, more akin to the eland but with huge outwardly curving horns. In vast numbers, ashen and ethereal, they churn through dust clouds or lie beneath their forest of horns, serenely aware of their special place in the order of things.

* * *

Christians are not so besotted with cows, but cattle are mentioned in the first book of the Old Testament and a calf in the last chapter of the New. The Bible teems with animals: in parables, as symbols and metaphors as well as in more lowly physical roles, ploughing and carrying and providing meat. Horses, camels, donkeys, goats and the ubiquitous sheep – the most common domestic animals of biblical times and terrain – are all there, but so too are less obvious creatures: lions and leopards, foxes and badgers, birds of prey and sea birds, ants and bees. Mostly it seems that these animals were noted rather than *seen*, evidence of God's omnipotence more than objects of wonder in their own right. And, in the divisive doctrine of Christianity, they served as metaphors for good or evil, clean or unclean, the saved or the damned: the 'white' sheep on one side, the 'black' goats on the other. There's not much place for the middle ground here, although the dappled and pied goats must have given pause for thought. Not much place for empathy either, but Proverbs has an endearing passage about 'four things which are little upon the earth, but they are exceeding wise', namely ants, conies (the rock hyrax), locusts and spiders – 'the spider taketh hold with her hands, and is in kings' palaces'.

The hyrax (Greek for shrew-mouse) is more closely related to elephants and manatees than to the small rodents it resembles. Its ancestry goes back some fifty million years, and this, the rock hyrax mentioned in the Bible, lives today in the Middle East and South Africa.

Christianity began at an uneasy time in the history of man's
relationship with the animal world: too late to enjoy the pre-agricultural
harmony that we imagine in the age of the hunter-gatherer, too early
to be concerned about the destruction to both animals and land caused
by the need and greed of modern humanity. For those living in Galilee
two thousand years ago, it must have appeared to be a land of ceaseless
plenty; though much of it was harsh and mountainous, it was blessed
with a high rainfall, and this terrain supported untold numbers of
grazing animals that provided meat, milk, skins and a livelihood for
the entire population. So perhaps the idea of the sacrificial lamb that
runs through the Old Testament a few thousand years earlier was not
as shocking then as it may seem to us today, though the slaughter of a
hundred bullocks, two hundred rams and four hundred lambs to mark
the dedication of a single 'house of God', as ordered by King Darius I,
must seem excessive to most people. The sacrificial lamb described by
the Old Testament prophet Isaiah, silent and accepting of its fate, would
become in the Christian story the Lamb of God. Unlike the scapegoat,
which was allowed to wander off into the wilderness, the lamb, bearing
the supposed sins of humanity upon its innocent back, is slaughtered
– surely a strange choice of symbol for a religion of compassion and
mercy.

* * *

For many of us, Buddhist or Christian, vegetarian or carnivore, animals
offer a Utopian vision of how we might live more simply, uncluttered
by the trappings of civilisation and materialism. Animals, both wild
and domesticated, have a purity of emotion that can seem enviable.
(Many times, after a fitful night in which minor problems loomed and
then intensified out of all proportion, I have greeted my dog who has
slept for ten hours solid without a care in his head, and wondered which
of us was the clever one.) Whether *seeing* animals helps us to attain
spiritual goals is questionable, but empathising with them – without
sentiment and with clear sight – certainly makes one feel calmer and
more grounded, more in touch with one's own nature as well as theirs.
Animals show us the pettiness of our troubles, in the same way that
facing death must do; we cannot take with us the wealth or the fame
we may have accumulated, we are reduced to our naked essence – the

state of animals at all times. These are surely spiritual lessons, even if not an exact religion. 'There was a time when I thought sweeter the howling of wolves, than the voice of a priest indoors, baaing and bleating,' wrote an anonymous Irishman in the twelfth century. In a time of much baaing and bleating, of politicians as well as of priests, we might do better to listen to the wolves howling – especially if they are howling for their planet.

2
DEPICTING
ANIMALS

One thing that even those who are most dismissive of animals have to accept is their omnipresence in the life and history of mankind. There they indisputably are, whether you like them or not, on the walls of the cave, at the foot of the knight's effigy, under the tables of the rich and in the farmyards of the poor, in myth and fairytale, in dreams and reality. The human need of animals, both practically and emotionally – even perhaps morally – is as strong today as it has been since we began to depict them all those thousands of years ago. Their domestication is a fascinating and often disturbing story in itself.

In most people's minds, the earliest visual evidence of our connection with animals comes from the now almost over-familiar European cave paintings, by no means the earliest examples of rock art. Yet such was their quality and sophistication, in form and execution, that when they were discovered they were initially believed to be fakes – in fact, the amateur archaeologist who discovered the paintings at Altamira in northern Spain in 1879, which besides being artistically some of the finest were also exceptionally well preserved, was at first suspected of himself forging them. Since then, over five hundred prehistoric painted cave sites have been confirmed in Europe, and more are doubtless awaiting discovery. Their purpose, however, remains a mystery.

Finding a 'purpose' for something as exuberant as many of these paintings, as lovingly and skilfully portrayed, is to me like trying to find a purpose for laughing, or living. Whether they were made as tallies by hunters, the most mundane possibility, or in order to evoke the animals, or as part of a ritual, or as what Kenneth Clark described as 'records of admiration', or just for the sheer pleasure of it, seems to matter little in the light of their energy, power and beauty. Whatever their cause and origin, these animals – horses, stags, bison, fleeing in herds or picked out in isolation, some in bas-relief on the rock face, others on specially flattened surfaces, in natural colours of black, russet, brown and grey – speak to us as vividly of their animal nature as when they were painted, 20-30,000 years ago.*

Although this may seem a very long time ago, in the timescale of evolution it is the mere passing of a bird in flight. One hundred and eighty thousand years were to pass between the estimated emergence of *Homo sapiens* and the execution of these accomplished paintings. During this time we slowly, very slowly, grew in intelligence, practical skills and language. So it is inevitable that throughout this time many tentative and transient attempts at artistic expression were also being made, in different parts of the world, in the sand, in mud, on wood and bark and leaves, using primitive tools and natural materials, as dexterity and curiosity began to distinguish us from other species of the *Homo* genus. About this early art we can only speculate, as it was of its essence ephemeral. But we know, for example, from a find in the Blombos Cave on the Southern Cape of South Africa, that an ochre pigment stored in shells, along with the tools used for preparing it, dates to the Middle Stone Age, between 100,000 and 70,000 BC. This allows a lot of time for experimentation, both practical and artistic, as skills improved and the imagination developed in early humans, alongside their relationship with animals.

For if these 'records of admiration' from the cave walls of Europe, Africa and Asia are any guide, man's artistic sensibilities and observational powers were inspired first and foremost by animals – not only those

---

* Dating of the paintings continues to be revised as technology advances and new discoveries are made. The caves of Lascaux in southern France and of Altamira in northern Spain, once thought to hold the earliest studies of animals, have now been preceded by sites in Asia.

familiar to us that provided sustenance and clothing for carnivorous man long before any attempts at their domestication, but wild animals such as the lion, rhinoceros and panther, which haunted his dreams and informed his religion. It is interesting, and a little strange, that at the dawn of representational art it was only animals that were portrayed realistically; the humans that did appear were crude, sticklike creatures.

Less well known than the animal paintings of southern Europe, though predating them by many thousands of years, are carvings from the Swabian Jura in south-western Germany. The Stadel lion-man, a hybrid figure with human body (whether male or female is still a subject of debate) and a lion's head, is claimed to be the oldest figurative sculpture yet discovered, made in about 40,000 BC. The upright figure, carved in mammoth ivory, is sophisticated not only in technique but in imagination. Carving in ivory with the use of a flint knife was a painstaking process – a modern reconstruction made with the same materials took 370 hours – so the figure was clearly important, almost certainly of religious significance. Equally skilfully carved, also in ivory, are tiny animals found in the nearby Vogelherd Cave. (It's a nice touch that this cave was discovered in the early 1930s only after some artefacts had been thrown up by a badger cleaning out its den.) The figurines, no more than 5-7 centimetres long, depict horse, mammoth, bison and what is thought to be a snow leopard. In each case the characteristics of these animals, their bulk and weight, are perfectly captured despite their minuscule size. They were made with obvious love, and could have been worn as pendants or sewn on to garments; either way their importance to their makers and wearers shines through.

These European paintings and carvings were made from direct observation by the Palaeolithic hunter-gatherers for whom the beasts were a part of life and a way of life; they were truly *seen* by

This woolly mammoth from the Vogelherd Cave would fit inside a small matchbox, but its power and dynamism are indisputable. It is intricately carved: even the soles of its feet are incised.

people who were clearly overawed by their power and ferocity. Lacking the animals' strength and speed, the early hunters also for the most part lacked the ability to outwit them. The odds lay with the animals. (Was this perhaps the reason for miniaturising them?) But by the time the next significant flowering of animal art came bursting forth, a dozen or so millennia later and scattered throughout different parts of the world, the wits and habits of the people had changed and with them their attitude to animals.

* * *

Although it was the cave paintings of southern Europe that first caught the public imagination, they were not an isolated phenomenon. For, as it has been at many stages in the long story of human development, similar things were happening simultaneously in places very far apart on the globe. In a cave on the Indonesian island of Sulawesi there's an extraordinary painting that has been dated precisely at 35,400 years old, though subsequent finds – and dating techniques – are likely to exceed its claim to be the oldest study of an animal in the world. It's a painting of a pig-deer, a babirusa, a member of the pig family native to this and a few nearby islands, still found there today. But what makes this image special is not just its age and unusual subject, but the way the artist has caught the characteristic physique of this least 'piggy' of pigs, with its tapered nose and delicate legs. This sow – only the males are mightily tusked – is portrayed by someone who was intimately familiar with the pig-deer.

The Australian Aborigines were painting animals – fish, tortoises, reptiles, wallabies – in their rock shelters at the same time that the cave-dwellers of southern Europe were painting the bison and horses we know so well. And although the subjects differed, the techniques were often strikingly similar. In the rugged landscape of the Kimberley, on the north-western tip of Australia, are caves with free-hand animal studies made with the greatest skill, the flowing brushstrokes of an outline filled in with solid colour while the body cavities are stippled and the extremities of fin or foot picked out in fine detail. The fish in particular are minutely observed, making their presence in a stony environment the more startling. Again, one wonders at the background of these lifelike studies: how the painter came to know his subjects, how long he sketched before committing himself to the wall, whether he had inherited his skill and would pass it on to his children and grandchildren.

The dot paintings most associated with Aboriginal art are heaving with animal life, the earliest examples made from direct observation and therefore varying according to the wildlife in different areas. These artists truly *saw* the animals; some actually depicted what was inside their subjects as well as what was visible. Much of this art is symbolic, less portrayal than a subtle language conveying practical messages: images for the tracks of different animals, reptiles and insects, the presence of water, the whereabouts of edible plants, all providing a survival map of the territory.

* * *

As human evolution continued its slow progress, all over the inhabited world hunting and gathering began to be replaced as a means of livelihood by agriculture and farming. In Egypt, by about 8,000 BC, the combined forces of over-grazing and climate change (does this have a familiar ring?) had driven the nomadic hunter-gatherers out of the Sahara and towards the fertile river Nile, where they settled and began to refine the process of animal domestication that had been slowly taking place, beginning with sheep in flocks – aided by dogs, bred from wolves – and going on to include cattle and pigs. This changed the relationship between man and these animals for ever.

But by now the association with other animals, some of them indisputably wild, had changed too, and what we have interpreted as the admiration shown by the cave painters had turned into adoration: animals as gods.

Animals held to be sacred, whether embodying gods or the souls of dead men or as mere metaphors, have been worshipped, sacrificed and depicted for as long as religion has existed, by the ancient Egyptians, the Bushmen of Africa, the Australian Aborigines, the Eskimos, the Celts and the Christians. The rock art of the San of southern Africa, already referred to, was more than mere representation or the paying of respect to animals that the people feared or hunted; it was itself a language rich in symbolism and metaphor: 'They put paint to rock and opened portals to the spirit world.'

The range of animals worshipped – birds and beasts, reptiles, fish and insects of all descriptions and sizes – takes this practice a long way from the few species of the cave painter's art. Nothing was too

This fanciful little creature, a mere 6.7 centimetres long, formed the top of a pin used as a fastener or simply as decoration. Dating from around the twelfth century BC, it is among many hundreds of small cast-bronze artefacts depicting animals found in Luristan, western Iran.

frightening, poisonous, odd or downright unattractive to become an object of veneration; some were possibly even chosen for these very qualities in the belief that if one befriended and honoured an animal, one would earn its protection. The balance of power was still in the animals' favour.

As part of worship there came representations: animal figures large and small, jewellery, amulets and talismans in a wide range of forms, styles and materials. From about 3,000 BC onwards, across Europe, Egypt, Africa and China, these artefacts poured forth as people of the Bronze Age began to display their skills, flex their artistic muscles and let their imaginations run free. The ability to work metals – gold, silver, copper, bronze – brought a new facility to the making of three-dimensional figures, both animal and human, and some of the earliest are both lifelike and engaging. Take the famous golden Maikop bull, one of a group in a similar style discovered in a burial mound in the northwest Caucasus. This alert and characterful little animal, made in the third millennium BC and a mere 7 centimetres high, manages to be both stylised and realistic, his muscular body dominated by the head with its exaggerated horns.

Chinese art abounds with animals both real and mythical, realistic and abstracted. The taotie was a stylised monster with one face and two bodies ending in coiled tails, often depicted in a flattened, linear design with florid embellishments. Early bronze ritual vessels from China were adorned, often encrusted, with interwoven animal forms – birds, snakes, insects – and there was an exact and complex animal symbolism: among the insects, for example, the cricket represented the fighting spirit, the grasshopper, wisdom, and the cicada, immortality.

* * *

The depiction of creatures that are half animal and half human goes back long before the mermaids of our own folklore, the minotaurs and satyrs of Greek myth, or even the Egyptian sphinxes. The oldest so far to have been discovered, the Stadel lion-man in the Swabian Jura already mentioned, was carved by skilled tool-makers of the Aurignacian culture, who also produced some of the earliest European cave art. A drawing dated around 30,000 BC from the Chauvet caves in southern France is a vivid close-up portrait of a figure whose lower body is that of a man and the upper that of a bison. Named 'The Sorcerer', the creature looms in front of an explicitly naked female form, 'The Venus'. It's a drawing in charcoal only, but the artist used the natural relief of the rock face to give an almost three-dimensional effect, and this plus the directness and physicality of this strange composition, which includes several other animals, add to its huge impact.

The combination of bull and man has aroused sexual fantasies ever since, embodying as it does the dichotomy that exists between the human and animal natures. The best known of these is probably the Minotaur, mythical offspring of Minos's wife, Pasiphaë, who had fallen in love with a bull. In the original myth the Minotaur has the head and tail of a bull and the body of a man, though most writers and artists have given him the head of a man on a bull's body, a rather more effective arrangement for his sexual prowess. The Minotaur makes an appearance in Dante's *Inferno*, and in his illustrations for the epic poem, William

Blake gives us an amiable-looking (and sexless) bull, more of a centaur than a minotaur despite the horned head and cloven hoofs. Picasso had no such inhibitions. True to his Spanish blood, Picasso was obsessed by bulls in general – he made countless drawings, prints and paintings of bullfights, many of them violent and gory – but his treatment of the Minotaur is exhaustive and much more psychologically telling, as the artist identifies himself with the contradictions of the half-human, half-animal figure in its aggressive sexuality and vulnerable pathos. In his great coloured etching of 1935, the *Minotauromachy*, it is hard to decide whether the approaching beast is friendly or threatening, beautiful or ugly – impossibly, it is both at the same time. Will it be tamed by the small girl clutching a candle and a posy of flowers? In an etching made a year before the *Minotauromachy*, the bull is blinded, groping with its head raised in anguish and led by the child.

Bulls recur throughout Picasso's work, from gentle beauties with curly hair to monstrous beasts intent on bodily harm. They are part of his mythology, and it is clear that they played a central part in his thinking. Few artists are as single-mindedly preoccupied with a particular animal as

William Blake, The Minotaur, *c. 1826, Dante's Inferno, canto XII.*

Picasso was, but animals of all kinds have padded, plodded and capered through artistic endeavour down the ages; even if only as extras, like the ox and ass in Nativity scenes, they are there, watchful and waiting, adding their colour and character to the often drab affairs of man.

Some stand out. Look at the sad, pensive dog sitting – his droop unforgettably expressive – at the feet of the dead nymph in Piero di Cosimo's *Satyr Mourning Over a Nymph*; or the elegant, thoughtful white hart with its golden coronet and chain from the fourteenth-century Wilton Diptych; or the mastin in the corner of Velázquez's *Las Meninas*, too happily comatose to be bothered to respond to the dwarf nudging him with a foot.

Many artists have included animals in their work without making them a priority. Henry Moore sculpted and drew a variety of beasts, both real and fantastic; his earliest sculpture, a curiously hunched dog with its paws on its nose, was carved in marble when he was in his early twenties, and he was still drawing animals, as frail as himself, shortly before he died aged eighty-eight. But it was in the realistic studies of the sheep grazing outside his studio, following their progress through a yearly cycle of birth, youth and maturity, from skinny to plump and from shaggy to shorn, that he really got inside the animal skin. Of animals in general he said, 'There can be a virility, a dignity or there can be tenderness, vulnerability,' and he captured these qualities in the improbable subject of sheep.

From the Renaissance, domestic animals have taken their place in serious works of art. Veronese was clearly a lover of dogs and often made them a focus of calm in his bustling, crowded interiors. Portraitists habitually use animals, dogs in particular, both as props and as a way of giving a lead-in to the character of the sitter. From Hogarth's pug embodying its master's own pugnacious spirit in his self-portrait, to Titian's Labrador visibly unmoved by the pose of its aristocratic owner (the dog would look just the same today, and equally unimpressed, at the feet of an owner wearing Hunters and a Barbour), to Lucian Freud's whippets, draped so intimately, yet at the same time uneasily, against the bodies of his subjects – all of these dogs add their own silent comment to the human business in progress.

Freud, who had come to England from his native Germany aged eleven and was a serious misfit in an English public school, found his comfort in the company of horses; he would creep off to the stables when life got too difficult, and even slept there from time to time. 'I feel a connection

Lucian Freud, Girl with a Kitten, *1947.*

to horses, all animals, almost beyond humans,' he said. He painted several pictures of horses as the sole subject, catching their character as he did with his human sitters, intent not to glamorise them but to expose their essence. He even dared to paint that most tricky of animals, a kitten, but the painting that shows his first wife, Kathleen (known as Kitty), staring into space with the little creature locked in a stranglehold and gazing at the viewer, is as far from kitsch as it is possible to get. The lack of visual contact between the two

subjects makes it a disturbing image, and the cat's awkwardly upright position is oddly reminiscent of an ancient Egyptian mummy.

For the most part, kittens and cuteness are synonymous and they get the exploitation they deserve, on chocolate boxes, calendars and YouTube. But there are more than a few serious felines that can hold their place in the history of art, none more so than an Egyptian bronze statue of Bastet, goddess of cats, dating from around 664-332 BC, which captures all the elegance and aloofness of these animals. The ancient Egyptians regarded cats with great reverence, and anyone killing a cat faced the death penalty. Cats owned by royalty were

A mummified cat, c. 30 BC, from Abydos, ancient Egypt.

treated royally, adorned with jewellery (like the golden ear- and nose-
rings and silvered amulet on the Bastet statue) and fed at the table from
their owners' plates. They were mummified in large numbers and offered
to the goddess – though this did dent their image somewhat, for it's hard
for even an Egyptian cat to look dignified when upright and swaddled
from the neck down, no matter how decorative the wrapping. When the
cat population of Bastet's temple got too large, as cat colonies tend to do,
the kittens were culled, mummified and sold to pilgrims.

* * *

Much of the earliest art, possibly as far back as cave painting, was
motivated by curiosity and a desire to understand more fully how things
worked. Leonardo da Vinci was insatiably curious, and he was the first
to make detailed anatomical drawings of the human body, delving
relentlessly beneath the skin in order to examine muscles, sinews and
bones. Not surprisingly his interest extended to animals, particularly
the horse, and we have many of his detailed and heavily annotated
sketches of horses that show their proportions and the workings of their
musculature, knowledge that he put to good use in many of his later
paintings. (Leonardo was lousy at drawing cats, though.)

It's worth remembering, in an age when the camera allows us to see
the innermost parts of a wild creature's hidden lair – and even of its
digestive tract – that such familiarity is a very recent privilege, and that
animals such as the elephant and rhinoceros, not to mention a vast array
of bird and insect life, were until recent times unknown to anyone who
lived outside their habitats. Albrecht Dürer, perhaps best known for his
painstakingly accurate yet vital paintings and etchings of animals – the
crouching hare is one of the finest – made a stab at depicting a rhinoceros
based only on a sketch and description he had received in a letter. The
famous result, equally painstaking but not quite so accurate, looks like
one of the Creator's wilder prototypes gone wrong, with its elaborately
decorated and frilled armour plating, scaly legs and platform hooves.

A hundred years later, by the early seventeenth century, European
artists became fascinated by the exotic animals being introduced,
often in travelling circuses, which they could observe at first hand.
Rembrandt's sketches of a benign-looking elephant are well known and
much reproduced, as is the recumbent lion – both far from the wild, and

Nach Chriftus geburt.1513.Jar.Adi.j.May. Hat man dem grofmechtigen Kunig von Portug. Emanuell gen Lyſabona pracht auß Jndia/ein ſollich lebendig Thier. Das nenen ſie Rhinocerus.Das iſt hye mit aller ſeiner geſtalt Abconderfet.Es hat ein farb wie ein geſprerkelte Schildkrot.Vnd iſt vō dicken Schalen vberlegt faſt feſt.Vnd iſt in der grōſ als der Helfande Aber nydertrechtiger von paynen/vnd faſt werhafftig.Es hat ein ſcharfſtarck Horn vorn auff der naſen/Das beginnet es allweg zu wetzen wo es bey ſtaynen iſt.Das boſig Thier iſt des Helffanten tode feyndt.Der Helffande furcht es faſt vbel/dann wo es Jn ankumbt/ſo laufft Jm das Thier mit dem kopff zwiſchen die fordern payn vnd reyſt den Helffande vnden am pauch auff vñ erwūrgt Jn/des mag er ſich nit erwern.Dann das Thier iſt alſo gewapent/das Jm der Helffande nichts kan thūn.Sie ſagen auch das der Rhynocerus Schnell/Freydig vnd Liſtig ſey.

Dürer's rhinoceros, drawn from hearsay and imagination.
The inscription delightfully describes the creature as being 'the colour of a speckled
tortoise, [with] a strong pointed horn on the tip of its nose, which it sharpens on stones.'

looking it. Later, wild animals kept in private collections were studied and included in made-up scenes such as George Stubbs's improbable *Lion Attacking a Horse*. Stubbs, master of horse painting, apparently observed the lion in someone's menagerie, but it's a sorry pussycat and it is the horse that visually dominates the painting despite its predicament.

* * *

Some animals just cry out to be sculpted or painted. Greyhounds, for example, with their pronounced body-shape and expressive faces. The greyhound is probably the oldest breed in the world, beloved by the Egyptians and, like their cats, often entombed with the Pharaohs. (It is, incidentally, the only breed of dog mentioned by name in the Bible.) There's a Roman marble sculpture named *Sick Greyhound* dating from

around 200 BC, from the ancient city of Gabii, that catches the essence
of greyhound – though whether this one is sick or merely repentant, I'm
not so sure; it certainly looks pretty dejected and has been brilliantly
realised. (The well-known Townley Greyhounds, subject of another
Roman marble sculpture carved two or three hundred years later and
now in the British Museum, are in fact thought to be salukis.) Dürer had
a go at a greyhound too, but not so successfully. Augustus John caught
the nervous intensity of the whippet, the small version of the greyhound,
to perfection, and Freud included them wherever he could, mostly asleep
on a bed.

Dogs could hog the limelight for ever, but perhaps the most painterly
– and arguably the most beautiful – of domesticated animals is the horse.
The subject of horses in art is worthy of a book on its own, and there are
many. From the sturdy, pot-bellied little animals of the Lascaux caves,
to the bounding, elongated war horses of the Egyptians and, later, the
Greeks; from the graceful studies of horses in motion that resulted from
Leonardo's anatomical probing to the contortions of frenzied energy
portrayed by Géricault and Delacroix; from the rounded, glowing
contours of Stubbs's horses in pastoral settings to the elegant racehorses
of Degas and Munnings – all of these varied treatments of a subject in
itself so varied are testament to the power of horses to move and inspire.
The horse represents the physical might of a wild animal brought fully
under control, 'harnessed' by man; no wonder then that it is such a
potent symbol.

Horses bridge the gap between truly domestic animals and the other
so-called domesticated animals, those that share our domain but not our
hearths, and rarely our hearts. In art as in life, the cows and sheep, pigs
and poultry are most often just *there*, foraging or grazing, pursuing their
own business as part of the landscape. Is this, I wonder, because they are
most likely to end up on the table so we don't encourage them to come
too close, or because they have not (yet?) learned to communicate with
humans as have dogs and, to a lesser extent, cats.

There are some exceptions, in both life and art. The potential for
communication between us and pigs is commonly boasted by their
keepers, but as far as I know no one has yet put up a case for having a
conversation with a cow. But that the Dutch painter Paulus Potter felt
for cows, observed them, knew them and loved them, is beyond doubt:
his unashamedly romanticised paintings catch the heavy contentment

PORTRAIT OF T.W.COKE ESQ? & CLERK HILLIARD ESQ?

This bloated and rather miserable-looking beast is obviously a source of great pride to its owner, Thomas Coke, 1st Earl of Leicester, who introduced lush grasses such as cocksfoot and lucerne to his estate in Norfolk in order to produce such specimens.

of bovines, their sweet breath and smooth hides, so that you can almost reach out and touch them, smell them. This is far from the case with the curiously rectangular representations of pedigree stock, mainly cows and sheep, that were in vogue in eighteenth-century England. These unconvincing, bloated creatures in their idyllic settings were drawn as portraits, and with much the same motives of self-aggrandisement on the part of their owners.

* * *

There is nothing new in animals being used to show off wealth and status, but more significantly they may provide pointers towards character traits or psychological states. A painter who used animals as alter egos in her work was Frida Kahlo; many of the self-portraits she made in a relentless record of her physical and emotional travails include monkeys, parrots, hummingbirds and butterflies. Particularly significant are the tiny spider monkeys with eerily human faces kept as pets and surrogate children by Kahlo and her husband, Diego Rivera. In several self-portraits these little creatures are at her shoulder with a hand around her neck; they are tender but somehow disturbing, an echo of all Kahlo's work. In the loneliness of her life as an invalid she found comfort in animals. As well as the monkeys and parrots of her native Mexico, she painted dogs – mostly the Xoloitzcuintle, or Mexican hairless dog – and some memorable cats. One of her most troubling pictures is of a chick dominated by a dark bouquet covered with webs and insects that seem to be threatening the little bird, overturning the usual order of things. The chick is taken to be Kahlo, who was burdened with depression and feelings of vulnerability at the time it was painted. But the most challenging self-portrait of all is *The Wounded Deer*. This extraordinary image of the artist as a centaur, her face instantly recognisable, with horns on her head and the realistic body of a male deer, presents a field day for psychologists. The hybrid creature (Kahlo when young often dressed as a man); the arrows, the only manmade thing in this natural setting, indicating both her bodily and emotional suffering (her love-life was turbulent); the dead-looking trees and severed branch in the foreground, images for her damaged body (her right leg was later amputated); even the perversity of the title, when the animal is clearly not a deer but a stag, is up for interpretation. But to me the strangest aspect of this very strange painting is that it has a serenity that belies its violent subject. Despite the suffering of both the animal and herself, the artist is at one with Nature, and with her own animal nature.

Animals make only rare appearances in the work of R.B. Kitaj. He drew cats – there's a beautifully observed study in pastel and charcoal of cats mating, coyly titled *My Cat and Her Husband* – but his painting of another animal is the stuff of nightmares. In *Jack My Hedgehog*, a human-sized monster with bared teeth – you can smell its fetid breath – more wild boar than hedgehog, its back-end fading into that of a man, straddles a white-clad, blood-spattered maiden. The supine, traumatised woman is reminiscent of some of Frida Kahlo's self-portraits (though for her, animals were a benign presence), the scene compelling: what is going on, we wonder? Kitaj's horrific image is in fact based on a fairytale with the same name and a happy ending, but you wouldn't guess it from this painting.

R.B. Kitaj, Jack My Hedgehog, *1991.*

So here from the end of the twentieth century is an image that takes us back full-circle to the drawing of some thirty millennia earlier on the walls of the Chauvet cave: the male as beast – perhaps predator, perhaps not, but in both cases physically half animal – and the vulnerable female. Whether presented as physical fact or metaphor, this merging of man and animal symbolises the opposites of tenderness and raw impulse that coexist within our own human and animal nature. For all our self-styled civilised humanity, we ourselves remain half animal.

3
DESCRIBING
ANIMALS

Visual images predated writing by many thousands of years, but since animals have been a vital part of the lives of humans from earliest times, it's not surprising that they also feature in early writings. Animals play a prominent part in various flood myths, from the familiar story of Noah's ark in the Old Testament to an even older, and similar, one in *The Epic of Gilgamesh*, the first recorded work of literature. People preparing to brave a great flood would not be so stupid as to leave behind their livestock; they took a male and a female of each to ensure survival of the species and a generous supply of sheep to keep the carnivores fed – what the herbivores lived on is not so clear, as no plants were included in the ark nor, for obvious reasons, fish. How the animals were perceived is not known either, whether they were *seen* or were there solely for their usefulness. But the fact that such effort was made to include all species, many of which would have been of no practical value, suggests that they were regarded as more than mere tools for man's survival.

Animals provided plentiful fodder for the fables attributed to Aesop, who lived in ancient Greece at some time in the sixth century BC, although his stories weren't recorded for several more centuries – and are still being added to even to the present day. They are morality tales, the moral spelled out each time, using animal characteristics to highlight human behaviour, particularly its faults and failings. Sayings such as 'a wolf in sheep's clothing', 'evil thoughts, like chickens, come home to roost', 'birds of a feather flock together' and 'fine feathers don't make fine birds' are all from Aesop's fables. Examining human conduct in the guise of animals had the added advantage of allowing Aesop, and others like him, to express subversive opinions without being openly critical: animals observed and in the service of humans from very early times.

By the time writing became widespread in Anglo-Saxon England, the animal population both wild and domesticated was much the same as it is today, with the addition of wolves, wild boar and beavers. Early medieval thinking held that there was a chain of being, a hierarchy with God at the top, then the angels, then human beings and lowest of all the animals; writers at that time comment on animals being hunted or used domestically, but little more. In the epic Old English poem *Beowulf*, the beasts came into the story as symbols of power and aids to dominance, along with some colourful monsters and dragons that had to be slain. Chaucer, in the mid-fourteenth century, used animals as metaphors for madness, lust and physical force. Even Shakespeare, that most shrewd observer of human traits, did not *see* animals in their essence but he used them frequently as pointers to human behaviour.

It wasn't until the later part of the eighteenth century that people began to write about animals as if they really saw them. Gilbert White, a country parson who lived for almost his whole life in the Hampshire village of Selborne, was an avid naturalist who combined acute observation with genuine feeling, particularly for the birds that were his special passion. These he logged almost daily over the decades he spent in this same village, raising questions and recording behaviour never previously noted. His records are meticulous, affectionate and often amusing, as for example his descriptions of the movements of different types of bird: 'Owls move in a buoyant manner, as if lighter than the air; they seem to want ballast. . . . [ravens] when they move from one place to another, frequently turn on their backs with a loud croak, and seem to be falling to the ground. When this odd gesture betides them, they are scratching themselves with one foot, and thus lose the centre of gravity. Rooks sometimes dive and tumble in a frolicsome manner; crows and [jack]daws swagger in their walk.' No one had recorded *seeing* animals in such a way before.

Fluidity and focus:
a lanner falcon encapsulates the opposing forces of the 'widening gyre'.

White died in 1793, when the Romantic poets were just beginning to find their voice. There's no doubt that Romanticism did romanticise animals – think of Keats's nightingale or Shelley's skylark – using them symbolically and often anthropomorphising shamelessly, but these poems surely made people stop and look at, and possibly *see*, animals for the first time. William Blake directly addresses those that are the subject of two of his best-known poems, 'The Lamb' and 'The Tyger', representing in their respective innocence and ferocity the opposing forces of good and evil. Robert Burns made clear his compassion for animals in 'To a Mouse', a cry of remorse after he turns up a mouse's nest while ploughing. Burns identifies with the mouse in its loss; he is down there with it in the field apologising for the wrecking of its 'wee-bit housie'.

Though the fashion for writing poems specifically about animals diminished after the Romantics, they have maintained their place in poetry ever since. Thomas Hardy leaned over a gate in the dusk and listened to a thrush 'fling his soul upon the growing gloom'. In 'The Windhover' Gerard Manley Hopkins famously rejoices at the 'dapple-dawn-drawn Falcon' whose

> . . . hurl and gliding
> Rebuffed the big wind. My heart in hiding
> Stirred for a bird, – the achieve of, the mastery of the thing!

How understandable that birds, bridging as they do the gap between the earthly and the spiritual, have inspired so much poetic imagery. The falcon again for W.B. Yeats, opening one of his best-known poems, 'The Second Coming':

> Turning and turning in the widening gyre
> The falcon cannot hear the falconer;
> Things fall apart; the centre cannot hold.

D.H. Lawrence wrote a series of poems about animals from direct observation when he was living in Sicily, and he wove references to animals through many others. He was no naturalist but an acute observer, who captured the essence of his chosen animal and painted vivid pictures for his reader. Of the mosquito:

> Queer, with your thin wings and your streaming legs,
> How you sail like a heron.

Of a fish:

> Your life a sluice of sensation along your sides,
> A flush at the flails of your fins, down the whorl of your tail,
> And water wetly on fire in the grates of your gills.

Lawrence isn't afraid to admit his fear or dislike of animals; his description of trying to drive a bat out of his room will strike a chord for many people, as the mutual panic mounts and he realises that the bat cannot leave the room because the light outside is too bright:

> . . . from a terror worse than me he flew past me
> Back into my room, and round, round, round in my room
> Clutch, cleave, stagger,
> Dropping about the air
> Getting tired. . . .

> Till he fell in a corner, palpitating, spent.
> And there, a clot, he squatted and looked at me.
> With sticking out, bead-berry eyes, black,

And improper derisive ears, and shut wings,
And brown, furry body. . . .

So, a dilemma!
He squatted there like something unclean.

Ted Hughes, who empathised with animals and wrote about them as
no one else has, also had a troubling encounter with a bat:

What looked like a slug, black, soft, wrinkled,
Was wrestling, somehow, with the fallen
Brown, crumpled lobe of a chestnut leaf.

Thinking it was sick, Hughes stooped to lift it to safety, when

It reared up on its elbows and snarled at me.
A raving hyena, the size of a sparrow,
Its whole face peeled in a snarl, fangs tiny.

Realising that the only way to catch it was to allow the bat to take hold of
his finger, to 'let the bite lock', Hughes does so and puts the bat back in the tree
from which it has fallen, then nurses his bloody finger. This is empathy indeed.

Hughes was a naturalist by experience, not by learning. As a child
he had what he described as 'a peculiar, obsessive relationship to wild
animals – simply their near presence. . . . It's a physical reaction: like
a kind of ecstasy.' His poems heave and pulsate with animal life and
extraordinary insights into their behaviour. From pig to crow to hawk
to fish, he brings his animals alive on the page in all their physicality,
imbuing them with symbolism and mythic force. Anyone who cares
about animals, who wants to *see* them without frills or sentimentality,
must read Hughes's animal poems.

* * *

If poets tend to focus on the symbolism of animals, writers of children's
books use them as communicators, natural bridges between the 'animal'
nature of the young and the 'civilised' world of adulthood. Animals, even
if caricatured, may act as interpreters for children, who are more open

than adults to the suspension of logic, more in tune with the simplicity of exchange that takes place between animals and humans. And through identifying with animals, children learn – about the animals, about the world, and about themselves.

The end of the nineteenth and beginning of the twentieth centuries brought a flurry of children's animal books, almost all of which have survived the test of time and retained their place as classics. The story of the horse Black Beauty has haunted children – particularly girls – ever since the book first appeared in 1877; it has sold over 50 million copies, has been made into a film nine times and adapted for the stage. It is a story of heartbreak, humiliation and redemption, a parable, social commentary and a tale of ultimate triumph over tragedy. It was also unusual for its time in coming 'from the horse's mouth', so that the reader identified with, in many cases *lived*, the ups and downs of the story. But it was not merely a tear-jerker for impressionable young girls. The book brought about a change in the law regarding the hours and conditions of work for horses drawing cabs, and the abolition – in Britain, though sadly not worldwide – of the use of bearing reins, which kept the driving horse's head up and stiffly elegant, at great cost to its natural poise and comfort. The book describes the plight and the feelings of working horses as never before; its author, Anna Sewell, herself an invalid following an accident as a fourteen-year-old, obviously empathised with the injuries suffered by Black Beauty, and she used this fellow feeling to introduce the idea of 'animal rights' to a public that had never thought this way before. She taught people to *see* the horses upon which they so heavily depended and had previously taken for granted.

Black Beauty: the covers may change but his spirit lives on.

Some twenty years after the publication of this ground-breaking story came Rudyard Kipling's *The Jungle Book*, very different in subject and setting but also out to make moral points through anthropomorphism. A series of tales, more fables than stories, are set in the Indian jungle and feature creatures unknown to most readers when the book was published – animals such as a cobra, muskrat, Bengal tiger and Indian mongoose, all friends of Mowgli, the jungle boy brought up by wolves. As well as providing a feast of animal lore, the stories are also seen as an allegory on the attitude of the British in India at that time to the 'natives', attitudes that mercifully have come a long way in the succeeding years.

In another huge scene-shift, the next author to capture her public by anthropomorphising animals was Beatrix Potter. A keen botanist and an expert on fungi, Potter was also a talented artist who illustrated her finds before producing her first book for children, *The Tale of Peter Rabbit*, published in 1902. Few children since have failed to be engrossed by the gently moralistic tales of rabbits, hedgehogs, mice and other small creatures, made human and fallible and very accessible by Potter's simple words and magical illustrations. She brought the animal world into the nursery, and made it a safe place to be.

Hot on the heels of Peter Rabbit and his friends came *The Wind in the Willows*, bringing the slightly more warty characters of Toad, Mole, Rat (actually a water vole) and Badger; the setting also a little wilder, a little closer to Nature. And here too, as in so many of the stories about animals for both children and adults of this time, the creatures are vehicles for human emotion – maybe the repressed emotions that people in a post-Victorian age were unable themselves to voice.

A few years later, in 1926, Winnie-the-Pooh made his first appearance. By then bears were losing the reputation for ferocity they held in many fairytales to become the cuddly teddies beloved by children at the beginning of the twentieth century, named after President Theodore Roosevelt who famously refused to kill a bear while on a hunting expedition. The bear in the first edition of A.A. Milne's classic, illustrated by E.H. Shepard, was very much a teddy but a dignified if slightly chubby one, though over the years many succeeding illustrators, ending up with Walt Disney, turned him into a pot-bellied object of ridicule. Which is a pity, because Pooh is an endearing character, kind and thoughtful if a little slow, a 'bear with little brain'. But this is anthropomorphism taken to new heights – or depths. The characters of the animals in the stories, toys of Christopher

Robin (Milne's son), have scant resemblance to the characteristics of the tiger, kangaroo, pig, donkey, owl and rabbit that they embody. Theirs is a cosy little world of fantasy, and for me it smacks of sentiment and exploitation. Animals deserve better treatment than this.

They began to get it with *Tarka the Otter*, published one year after *Winnie-the-Pooh*. Aimed at an older age group, *Tarka* soon attracted an avid adult readership, and revisiting it myself sixty-odd years after my first reading, I found it easy to see why. For quite apart from identifying with the otters as characters and following their fortunes – in a real setting, along the Devon rivers of Taw and Torridge – one is swept up in an environment so credible and so vividly described that the whole experience is as if one were there with the otters, smelling what they smell, seeing what they see, playing and foraging and exploring their world. To read of Tarka's first encounter with water is to re-experience one's own: 'he dropped down into the black, star-shivery water. He was clutched in a cold and terrible embrace, so that he could neither see nor breathe, and although he tried to walk, it smothered him, choked him, roared in his ears'.

This was a new way of writing about animals – the equivalent, I realised, of the experience available to us now with the latest camera technology: one is 'on the animal's back', seeing the world from its viewpoint, down there with the otter pushing through the reeds or cowering in his holt as the huntsmen pound it with iron bars. Yet Williamson's writing is anthropomorphism in a different guise, for however closely he watched otters – and he did so obsessively for four years while writing the book – he attributes to them feelings that bridge the gap between animal and human. That he backs up these flights of the imagination with facts about the natural world keeps the narrative from becoming too fanciful. For example, he justifies Tarka's reaction to his first immersion – 'The first otter to go into deep water had felt the same fear that Tarka felt that night' – by explaining that as members of the weasel family along with badgers, stoats and pine martens, otters were once land animals, having only recently, in evolutionary terms, taken to the water. However lyrical his descriptions, however closely we engage with the otters as personalities, they remain real animals, living an animal existence in a natural setting and in constant fear for their lives. For Williamson does not spare the blood and violence of life in the wild – in fact he has been accused of revelling in it – and as the hunt closes in towards the end of Tarka's short life we face the otter's death with him, and suffer.

(Williamson's romantic view of otters has recently been debunked by Charles Foster, who goes one further than the camera on the back by attempting himself to *be* an otter, as well as a badger, fox, red deer and swift, by getting into the water, down on the ground and up into the air with each creature. That he fails utterly in this crazy endeavour, as he cheerfully acknowledges, doesn't lessen the impact of what he has to say, and I for one will never again feel readily sentimental about otters, whom Foster describes as 'these primordial stoats. . . . these jangling, snarling, roaming, twitching bundles of ADHD'.)

* * *

The differing realities of Williamson and Foster are a far cry from the twee world of Pooh Bear in the mid-1920s, though the run of popularity for the bear as entertaining character was far from over by then. Rupert Bear started life as a comic strip in the *Daily Express* at the beginning of the twenties, but it was not until 1935 and a new illustrator that he really took off – and has never looked back. The *Rupert Annual* has been published every year since 1936, even during the paper shortage of the Second World War; there have long been Rupert Bear fan clubs, and he now has his own website that boasts 'official followers'. He retains his yellow checked trousers and scarf and red jumper, though his chunky brown boots have been replaced by red trainers. It is hard to say what accounts for his enduring popularity; he's an innocuous character – polite, obedient, wanting to please, a model son to loving parents, not the sort that children usually identify with.

Paddington Bear, a later addition to the speaking bear family, is also polite and kind hearted though he has a more exotic background than Rupert, having come from Peru where he had been orphaned in an earthquake. But both Paddington and Rupert are unashamedly designed to be cute, and neither has anything whatever to do with bears as animals, nor are they *seen* as animals; dressing animals in human clothing immediately takes away their 'animalness' as well as their dignity, and Beatrix Potter has much to answer for in starting the fashion for dressing up animals.

Coincidentally arriving on the scene in the same year as the first Potter book, the original teddy bears were bare and relatively bearlike, though they were soon to be clothed by storytellers and salesmen. They also started off with the small eyes and long snouts of real bears, but over time their faces were squashed and their eyes enlarged to make them look

From bear . . . to teddy bear . . .
to teddy.

less bearlike, more human. Quite why it was thought necessary to clothe teddies is unclear, as bears – like most animals until they get going – have discreetly hidden genitals that couldn't possibly give offence to even the most uptight of Edwardian nannies. But maybe blurring the edges between the physicality of an animal, especially one as wild as a bear, was inevitable in an age unable to accept sexuality as natural but something to be covered up, if not denied altogether.

People have been clothed by animals since prehistory – and still choose to be, though the fur will probably now be fake – but dressing animals is a modern phenomenon. And as soon as they are clothed, the animals take on human characteristics. They need, for one thing, to stand on two feet, and as we know from *Animal Farm*, this is the beginning of their moral decline. The moment when the pigs first emerge from the farmhouse on their hind legs, closely followed by their leader, Napoleon, carrying a whip in his trotter, was 'as though the world had turned upside-down'. The first of the Seven Commandments that marked the rebellion of the farm animals, 'Whatever goes upon two legs is an enemy', had been broken. The seven commandments were later reduced to the single, infamous 'ALL ANIMALS ARE EQUAL BUT SOME ANIMALS ARE MORE EQUAL THAN OTHERS'. And from then on the pigs dress up in the farmers' clothes, 'Napoleon himself appearing in a black coat, rat-catcher breeches and leather leggings, while his favourite sow appeared in the watered silk dress which Mrs Jones had been used to wear on Sundays.' There is no need of an illustrator to get this scene across; the picture is vivid, its message all too clear.

Animal Farm was published in 1945, but writers had begun to use animals in adult fiction a century earlier. *Moby-Dick* was a flop when it was first published in 1851 – it was described as an 'ill-compounded mixture of romance and matter-of-fact' in one English review – and didn't take off for another fifty-odd years. Melville was aware of the difficulty of writing a successful novel based on whaling: 'It will be a strange sort of book, tho', I fear; blubber is blubber you know; tho' you may get oil out of it, the poetry runs as hard as sap from a frozen maple tree.' But he did succeed in finding poetry in this allegory of the struggle between good and evil, later described as 'a profound, poetic inquiry into character, faith and the nature of perception'. Melville's observations of whales and whaling were vivid and first hand; he himself spent several years on whaling ships, and the protagonist for his book was a real sperm whale, Mocha Dick, a renowned albino that lived in the South Pacific during the 1840s.

The theme of the battle between man and beast – and the sea – was taken up by Ernest Hemingway in *The Old Man and the Sea*. Here it is a single-handed fight between the old fisherman, Santiago, and the fish he desperately needs after a long period of bad luck without a catch. The marlin he finally hooks is a monster, and in the two-day battle that ensues Santiago comes, through his own ageing body and his pain, to identify with the fish, to 'think like a fish' and to honour its bravery and suffering. In this there is no victory, for the much-prized marlin is eaten to a skeleton by sharks as it is being towed home, and the old man retires to his bed to dream not of fish but of lions in Africa.

Animals occur in everyone's dreams, and it is perhaps significant, to psychologists at least, that the dream animals of children under five are not predominantly the small, cuddly sort one would imagine but large ones such as horses, lions and bears. One of the most disturbing dreams in literature comes in *Crime and Punishment*. The book's leading character, Raskolnikov, who has planned the wanton murder of an old woman, has a dream in which he watches a horse being savagely beaten to death by a gang of drunks. The scene is horrific and graphically described, and Raskolnikov, who is only seven years old in the dream, feels great compassion for the horse, though this doesn't prevent him carrying out the murder. Many layers of meaning have been attributed to the dream beyond the obvious parallel between the innocent victims, the horse and the woman he eventually murders, among them that the horse represents Mother Russia. Such are the burdens that animals carry for us, in dreams as in life.

* * *

This glance at the colourful parade of animals through literature would not be complete without mentioning two very different books that appeared within a couple of years of each other in the early 1970s. *Jonathan Livingston Seagull* is a barely disguised allegory for the limitations of modern life and the desire of the soul to reach – in this case physically – a higher level. Jonathan's disillusion with materialism and search for his 'true self' reflect the spirit of the time, and many young people identified with the bird's passion for flight, a symbol of spiritual freedom.

Similarly anthropomorphised, though with their feet very solidly on the ground, are the rabbits in *Watership Down*, published two years after *Jonathan* in 1972 and aimed at a younger audience. This is more a straight adventure story than a moral tale, though inevitably some try to read more into it despite Richard Adams's assurance that he never intended it to be 'some sort of allegory or parable'. It's a new take on the animals-as-humans story, as the rabbits live in their own environment, are not clothed, but have their own language, Lapine, and a culture rooted in myth and poetry. And like so many other ultimately successful children's books, it sprang from tales told to the author's own children, and was rejected by a host of publishers before becoming a bestseller.

All the books considered here add something to the sum of our knowledge of animals, help us to *see* them even if only as caricatures of their true animal nature. Favourite animals such as Black Beauty, Peter Rabbit and his friends, Winnie-the-Pooh and his, the riverside inhabitants of *The Wind in the Willows* are embedded in the psyche of most Westerners, and despite their different illustrators most of us will have our own visual images of them: they are part of us and helped to inform our childhood.

* * *

Visual imagery must surely be responsible for the phenomenal success of two theatre productions centred on animals: the musical *Cats* and the stage play *War Horse*. Who could have predicted that poems about cats written by a serious poet – T.S. Eliot, who first wrote them for his godchildren – and a children's story about a horse in the First World War would become world-famous, record-breaking hits in the West End and on Broadway?

Both demanded huge leaps of imagination to turn written descriptions into visually exciting performances, and in both cases failure was predicted, particularly in the case of *Cats* that remains faithful to Eliot's verse in *Old Possum's Book of Practical Cats*. Eliot died sixteen years before the musical's first performance in 1981 so he had no say in the matter, but for Michael Morpurgo, author of *War Horse*, steady nerves were demanded. 'When I first heard that they were going to make my story a bunch of puppets on the stage, I had no faith at all.' The success of both shows, so different in subject and treatment, is of course down to the talent and creativity of their makers, but these drew on their knowledge of the animal protagonists for inspiration: the explosive energy of the cats as they fill the theatre in an orgy of dancing and singing, and the qualities of loyalty and courage – combined with great looks – that define Joey, the horse who is both hero and narrator of *War Horse*. Both representations engage with our conceptions of these different animals on a deep level; as one critic said of the latter, 'I would wager that for a good while, you'll continue to see Joey in your dreams.'

In our dreams as in the real world, on the page and on the stage, these animals stir our imagination while they keep us in touch with their reality, and through them we come a little nearer to keeping in touch with our own.

4
MEETING
ANIMALS

Meeting animals, which entails *seeing* them and possibly communicating with them, is a haphazard business. In the wild there are no rules, and any control imposed on domesticated animals and those in captivity is open to disruption at all times. This is, to my mind, one of the most engaging aspects of relating to animals. As with other forces of Nature, we humans are put in our place as observers and participants, not masters.

Animals are not predictable. We can study their movements, make assumptions about their habits, track them with modern gadgets as our forebears tracked them on foot, but we cannot tell with certainty which way they will go, or when, or even whether. Advanced computer technology and artificial intelligence are now being employed to take the 'tedium' out of observing animal behaviour – largely, it must be said, so that humans can laboriously learn what the creatures do instinctively – but even the most sophisticated technology merely enables us to wait and watch, without being able to influence what takes place.

The 'Diaries' section at the end of the *Planet Earth* series makes this all too clear. The seemingly effortless encounters of the main programmes, with startling close-ups of previously scarcely known creatures great and small, every hair and whisker, every quiver and convolution, so close we feel we can touch and smell them – these come at a price. Dedicated teams plan, organise, kit out, transport and feed the people who 'hunt' the perfect picture, spending days and nights in cramped hides, too hot, too cold, prey to every biting bug known to man – and a few new ones to delight the entomologists among them. And often, all is in vain: the animals just don't show.

It's a similar story at home. No walk through a certain stretch of countryside, even around an urban park, is predictable however often you make it. The odds are you will see some sparrows and blackbirds along the hedgerows, or dogs and ducks and maybe a few grey squirrels in the park, but more memorable sightings come when least expected. One day a stoat perhaps, but never again in that place; or a hare leaping from its hide – you wouldn't have seen it there – and zigzagging off across the plough; in the fading light, the ghostly flicker of a barn owl. Here in Brittany, once only in seven years, I watched a family of red squirrels doing a high-wire act, mewling with excitement (or it seemed to me like joy) as they hurled themselves again and again from top to bottom and from side to side of a forty-metre Douglas fir. Even in London you may catch sight of a kestrel hovering among the City skyscrapers, or a salmon in the river Thames. But none of these sightings can be arranged; they happen, or they don't, and most people are too busy going about their own lives to notice the animals going about theirs. For anyone open to these brief encounters, they bring a frisson of excitement and leave behind a glow of wellbeing.

There are heart-pumping moments too, equally unpredictable whether in the wild or at home: the unstoppable fury of a dog fight, a rearing horse, donkeys on the loose heading at full pelt for the main road. As I lay dozing one late summer's day on an overgrown lawn in Hampshire, I was vaguely aware of movement beneath one hand. Suddenly alert, I sat up to see a skein of maybe five or six tiny adders slither into a nearby tree stump. On a hillside of scrub and holm oak in Andalusia, I sat against a rock while my dogs rooted around in the undergrowth behind me. Idly daydreaming, I remained oblivious to a more serious noise – until there erupted beside me a massive, and massively tusked, wild boar. He stopped, turned and looked at me, and for a second – it seemed much more – we were eyeball to eyeball, his heart, I imagine, going as fast as mine. Then the moment broke; he turned and crashed back into the scrub, and I breathed again.

'Pretty and plucky'.
The muntjac,
oldest of the deer
species, comes from
South Asia but has
settled happily – if
controversially – in
British woodland.

Richard Mabey describes a more dignified meeting he had with a muntjac, the tiny feral deer now common in southern England after escaping from captivity. This was not a sudden flustered encounter but a slow sidling that ended with the two staring at each other from about three metres apart. 'She looked at my eyes and passed her tongue repeatedly over her face, wondering if I was dangerous or musty. I thought she was pretty and plucky. She thought I was odorous and interesting and curiously shaped and things I could never know. We conversed in this way, two curious and diffident strangers, then went our separate ways.'

* * *

Communication must by definition be a two-way process, and I think it must include a degree of engagement on both sides. But that doesn't mean that the one we are communicating with must reply in kind. That we can communicate with animals in many ways that preclude, though may also include, speech is indisputable. And to those familiar and in sympathy with animals, their non-verbal response is clearly understandable. As the parrot informs Doctor Dolittle, who has given up on humans and is embarking on his career as an animal doctor, 'Animals don't always speak with their mouths. They talk with their ears, with their feet, with their tails – with everything.'

The fictional Doctor Dolittle learns the language of animals, but when we see David Attenborough talking to a salamander perched on his hand, we do not expect the creature to answer. However, I hold that the look on Attenborough's face as he talks to and about the salamander is of a different order to the look he would have when examining a flower. There is something about a sentient being that calls forth empathy; we are *seeing* the creature, and it is allowing itself to be seen. A communication takes place in which eye contact, body language and aura are all likely to be involved. But there is more than this. Michael Morpurgo, who besides being the author of *War Horse* also runs a charity called Farms for City Children, tells how Billy, a boy who was largely mute due to a crippling stammer, is found in the stables after a few days on the farm, 'Talking, talking, talking, to the horse. . . . she knew that she had to stay there whilst this went on, because this kid wanted to talk, and the horse wanted to listen – this was a two way thing.' An urgent, pressing communication, and who knows what the horse perceived that we humans could not. As Russell Hoban puts it, 'Any horse you meet seems to have in it a knowing that is deeper, older, more primal than our own; they seem to witness something lost to the sight of humankind.'

Non-verbal communication has a rich vocabulary, and many of the noises we make when showing emotion are shared by the higher animals. Screams, grunts, moans, sighs and whimpers don't need to be learned; these 'animal noises' are instinctive and, unlike body language, common to people of all races. And it is because the emotions can be expressed without words that animals are safe repositories for our confidences and fears. They may not be able to offer advice but nor do they judge, and we know that all secrets are safe with them. They see through any falsities and pretensions and expose us for what we really are – not always a comfortable experience. But these mirrors are not the cold, hard rectangles of coated glass that reflect our hangovers and declining years but rather a warm, forgiving being in whose presence we can truly express ourselves.

* * *

'In the very earliest time, when both people and animals lived on earth, a person could become an animal if he wanted to and an animal could become a human being. Sometimes they were people and sometimes animals and there was no difference. All spoke the same language.' No one knows when 'the very earliest time' was, but these 'Magic Words' from an unknown Inuit poet speak for themselves. In the twenty-first century we no longer speak the same language – if we ever did – but I believe we still yearn to do so.

To the sceptical, such claims of speaking the language of animals sound like nothing more than sentimentality, or senility: the old woman talking to her dog because she has no one else to talk to, poor thing. But for those two it is indeed a conversation, with subtleties of expression and nuances known only to the participants. It is a rich language, verbal, visual, tactile and emotional, one whose vocabulary increases over time and is unique, to be learned and developed anew with each new relationship.

It's a long journey, learning each other's language. Consider the small puppy, away from its mother for the first time, in the arms of a stranger on its way to a new home and a new life. It knows nothing but the smell of the person who holds it – and the smell is not just of jacket, of skin, of tobacco or cosmetics but of aura: of confidence and calmness if it is lucky, of panic and stress if not. Physical smell plays a part for the person too; maybe this scrap in your arms comes across as a bit doggy, almost rank, but the essential smell is there, a unique smell that you imprint on just as the dog imprints on yours. The bond is formed, and from this the communication will evolve.

We all know instinctively, from an early age and without being told, that physical contact with animals, stroking in particular, is calming and restorative for people (and, mostly, for the recipients too). But this now has scientific backing from a study showing that petting an animal releases oxytocin, the hormone associated with the pleasure felt during other tactile activities such as breastfeeding and love-making. Certainly many people keep animals as surrogate children and/or lovers, either because they have never known 'the real thing' or because it is no longer available.

As with a relationship between people, the link between animal and human has to be worked at or it will stagnate. There is always more to learn, on both sides, and life will continually throw up opportunities for taking a further step.

There are of course different expectations, depending on the animal. Man's communication with dogs is older than with any other animal, and arguably the closest, but even that most aloof of creatures, the cat, has learned how to combine total autonomy with the comforts of regular food and a warm hearth – and go for them, as only cats know how. No one in their right mind will attempt to train a cat as they would a dog, and the conversations between the two species and humans are of a different order. But perhaps it is the cat's very containment that makes it a solace to people in the borderland between sanity – our modern view of it, that is – and madness. Jay Griffiths, groping her way in that half world, writes of cats: 'They are an exercise in instant mindfulness. . . . They cannot but live in an eternal present and do so beguilingly, drawing us, too, towards the glow at the heart of now.' Griffiths quotes the eighteenth-century poet Christopher Smart, whose cat, Jeoffry, was his companion during six years spent, probably wrongfully, in a mental asylum. Smart's poem to Jeoffry is a masterpiece of observation and a testament of devotion, for without a cat, he writes, 'a blessing is lacking in the spirit'. Cats mend minds not because they are trained to do so, but simply by being what they are; they heal because they themselves are whole.

* * *

Cats are not my favourite animals. I like them because they *are* animals, and I respect their independence and am interested in their behaviour. I have provided a home, but not always a hearth, for several cats in my life, two of which I have loved even while still not really liking them; but with none of them has there been what I would describe as communication. The relationship has been strictly on their terms.

The story of Delilah demonstrates this well. She came to us young and semi-wild – a mouser was needed in a newly acquired old house – and because my two hefty dogs were not used to cats, I realised that some slow introductions would be necessary. So Delilah was put in a large, high living-room (in the process of being restored) so that she would, I hoped, get used to seeing the dogs around, and vice versa. But before introductions could be made she had disappeared. No sign anywhere; behind the built-in cupboards and the wood-burner, up the chimney, under the furniture, all were scanned by torchlight, but no Delilah. There were no sightings the next day, despite offerings of food, water and several uninterrupted hours when she might have dared to emerge. But the following morning the food had gone. This continued for several weeks, during which time we guessed that she had found a way up behind some panelling to the gap between the ceiling and the floor above.

When at last this panelling had to be removed, there were signs of Delilah's existence but not of her. Thinking that she must have somehow fled unnoticed in all the confusion, that night I put food outside (well protected from the dogs), and the following day it had gone. For the next eighteen months I continued to feed Delilah this way every night, and during this time I saw her only twice, fleeting glimpses in the twilight. Then one evening, as I went out at the usual time with her food, a shape appeared from the bushes, strolled across to me, and wound itself around my legs.

Eventually she became quite tame, affectionate even in the stand-offish way peculiar to cats. I have often wondered what went on in that cat mind, with no human contact for more than a year and a half, without any history of human contact, that she should suddenly approach me like that, for no apparent reason – not fearfully or hesitantly but purposefully, as if she had been doing so all her life, or had worked something out. This is why I don't understand cats, but continue to be fascinated by them.

* * *

While dogs communicate with us actively – by eye contact, different types of bark, pushing with nose or paw – and cats make it all too clear what they want and expect, the other domestic animals tend to be more reactive in the way they respond to human company. Even the horse, perhaps the next of our close domestic animals to be able to make its needs known to us, one might think can only whinny, nicker, and show its pleasure at your presence by a limited body language. But recent research has claimed that horses behaved differently when presented with pictures of smiling or angry faces: when shown angry faces, their heart rate went up and they turned to look with their left eye, a reaction associated with perceiving negative stimuli, but they gave no adverse response to the happy faces. So horses too *see* more than we may imagine.

Where the horse is unique among animals is in its response to the physical presence of the human on its back. And presence is the word. Not the mere physical bulk of the person but his or her vibe, the intangible quality that surrounds them as they come into contact with the animal.

It is perhaps easiest to think of this as a person's aura – one to which we too, as humans, are susceptible – that is not to do with behaviour or actions, though it affects both, but with essence. For it is a quality that cannot be simulated, and is perceived intuitively by both animals and other humans. It is what we are, our quiddity, the psychic equivalent of our physical smell.

Although there may be a strong communication between the horse and whoever looks after its daily needs, the rapport that can develop between horse and rider is something entirely different, and truly remarkable. Here is a human being seated on top of a large animal, out of eye contact, combining a limited verbal language with only the subtleties of touch, pressure and balance to persuade the horse to perform an extraordinarily wide range of tasks and movements. Of course it is possible to sit on a horse, wave the reins and flap your legs and you will get somewhere, though most probably not where you intended. But developing a genuine language between horse and rider is a serious undertaking, complex and slow to develop, demanding patience, dedication and trust on both sides.

Someone who has both studied the theory of advanced horsemanship and put it into action is Vicki Hearne, who besides training horses and dogs is also a philosopher. Her books are not easy, but anyone interested in the discourse between horses (and dogs) and humans will learn how intense the relationship can be, at both ends of the spectrum of frustration and satisfaction. This is communication at the highest level, and it is not for wimps. The horse and rider, with mutual trust, 'converse' – Hearne is good at describing these conversations. The horse gives of its instinctive knowledge, the rider contributes his or her analytical observation, and the two come together to produce the required result: what Hearne calls 'participation in knowledge'. This clearly does away with the idea of the person being dominant; rather, it's a status quo built on the animal's trust and acceptance of the rider's control, a trust that has to be both earned and honoured.

As a child I always wanted to be around horses, and chose to spend my spare time mucking out, grooming and occasionally, oh joy, exercising the horses in a stables near – but not so near – where I lived. Long bicycle rides on frosty mornings, hanging around all day in bitter cold with wet feet, heaving barrowloads of manure across a soggy yard – none of this dampened my enthusiasm. But there was no one there to teach me what real horsemanship was, or how to communicate with the animals on anything but a very basic level.

So when, fifty-odd years later, I decided to keep a donkey, the real equine education began. First I had to learn that donkeys are not horses. Donkeys don't set out to please, whereas most horses do. Because of this, donkeys are labelled stubborn, but it is more that they know what they want, and when, and how to get it – which is by ignoring any human attempt at control or intervention. They are also slow, even by animal standards, both to learn and to react. So the person who takes on a donkey, and does not resort to the (futile) use of force, has to learn to go slowly too, to persuade by encouragement and repetition, to coerce, and to be very, very patient. But as the trust builds, so does the communication.

Like most four-footed creatures, donkeys respond to the human voice. And the voice, of course, reflects the aura of the person concerned; donkeys will become attuned to the timbre of their handler's voice, if not the words used, and react accordingly – or not, if the wind is in the wrong direction, for donkeys are never predictable. But despite their reputation for contrariness, they are not untrainable: it is claimed that teamsters, large donkeys imported into Australia from Spain in the nineteenth century and then crossbred to produce particularly strong working animals for opening up the land, were driven in teams of as many as thirty by voice alone.

Donkeys are renowned for their own voice, the unmistakable bray that seems to be dredged painfully from their entrails and can be heard as much as three kilometres away. They have other noises – a whicker not unlike that of the horse and an excited heavy breathing when food is anticipated – but most of the communication from their side consists in action, a basic body language of approach, nuzzle, push, and general appreciation of handling and affection. It's a limited language but one that gives mutual pleasure, and it grows over time; from donkeys one learns never to be in a hurry.

* * *

In that slowest timescale of all, evolution, familiarity with humans has increased the vocabulary of domesticated animals. And for those with the patience to teach them, some animals show an astonishing ability to learn our language. There have been many claims for horses being able to solve simple mathematical problems by tapping their hooves, perhaps the most famous being an Orlov Trotter called Clever Hans, who became a celebrity for this so-called skill in the early 1900s. He was discredited when it was

Clever Hans with his owner,
Wilhelm von Osten, and performing
to an audience, 1904.

proved that he was not doing the maths himself but merely responding to the minute and unconscious signs given by his handler whenever the right answer was imminent. This seems to me even more clever than the ability to do sums, but it was not the sort of intelligence that people had been hoping to detect.

But interest was aroused, and soon afterwards there began a series of experiments in animal cognition in which long-suffering animals, usually under laboratory conditions, were subjected to exhaustive trials, puzzles and mazes dreamed up by enthusiastic behavioural scientists. Not my idea of fun, or even of fair play, and certainly far removed from natural conditions, but humans will go to great lengths to prove their theories. Quite what is proven also seems to me somewhat specious, and best left to those scientists and philosophers who, like kittens, enjoy chasing their own tails.

Closer to natural conditions, and far more like fun, is a border collie called Betsy who recognises about four hundred words and is able to identify objects from coloured reproductions on paper, then go off to another room and find them. Border collies are reputed to be the most intelligent dogs in existence, this presumably having evolved from their use as working sheepdogs and their consequent understanding of verbal and physical commands. They are quick dogs, eager to please and pleasingly eager: Betsy at work is a very happy dog.

But for most of us such dedication is not on the agenda, and communication with animals, whether in the home or the farmyard or in the wild, is a more casual, haphazard affair. For me, gradually finding the essence of the animal that shares, in whatever degree of intimacy, my life is a more achievable and more rewarding aim. In this journey of mutual discovery we find out not only about the whims and foibles, pluses and minuses, of the dog, cat, donkey, bird or whatever animal it is that turns up in our lives, but about our own.

* * *

An animal that is under-reported in terms of its intelligence and ability to communicate is the pig. The most celebrated pigs are the leaders of the rebellion in *Animal Farm*, 'manifestly cleverer than the other animals'. Orwell himself cleverly catches the characteristics of the various farm animals, though they are caricatures and we go along with them as the allegory that was intended. But in *The Whole Hog*, Lyall Watson gives an account of very real pigs – pigs from all over the world, from the bush pigs of his childhood in Africa to the babirusa of Sulawesi we came across in Chapter 2. Watson, a highly qualified zoologist and biologist, is much too serious to fall into any of the possible traps of sentimentality or anthropomorphism, but he does claim to have had 'close relationships' with three species of wild pigs, and writes with huge affection for the individuals he has known and for their genus worldwide.

Watson quotes Winston Churchill, another notorious admirer of pigs, in typically perceptive form: 'Cats look down on you; dogs look up to you; but pigs look you in the eye as equals.' This habit of staring one in the eye, uncommon in the animal kingdom, can make one feel uncomfortable among pigs: being *seen* by animals means being seen as we are.

But this is the serious side of pigs, and there is another. They love to play, and most people who keep pigs have stories of their antics and many claim for them a sense of humour. There is a jollity about their behaviour that is lacking in, say, a sheep or a cow. They are companionable, with each other and with humans, and have much in common with us physiologically; in terms of internal organs, though not of DNA, pigs are our closest cousins. Consider as well the externals. Their pink flesh, Beryl Cook chins, rounded buttocks and high-heeled trotters don't need the addition of a flowered hat to make pigs look all too human - and I am sure they must also be bawdy.

Which makes it seem strange that they get such a bad press. The very word 'pig' invites pejorative, and largely undeserved, adjectives such as dirty, lazy, greedy, and add to that swine, hog and porker and you have a fine armoury of abuse. Yet the intelligence is evident. It may come as a surprise to those who think of a pig's language as consisting mainly of *oink* with the occasional squeal that they do in fact have a rich variety of meaningful sounds, communicating vocally beyond the mere passing on of information. (In *The Whole Hog*'s index, the word 'communication' has more sub-entries than any other.) What we in our ignorance label simply a grunt can have many meanings and nuances. There's research currently investigating the interpretation of their different sounds, both in the wild and, more importantly, in controlled environments, where better understanding of pig language could lead to their being treated more humanely during their often tragically short lives between birth and bacon.

* * *

However much we may disparage the sounds made by pigs, few would argue about the beauty of birdsong. Why do birds sing? It's an old question, and it is their song that brings us closest to the birds, something we share with them, though our motives for singing are rather different. For most birds, their song is a form of communication, but one mostly designed to display sexual attractiveness on the part of the male. They make a wide range of other noises of course, to define territory or give warning of danger, calls that bird books strive to transliterate in bewildering ways such as *tee-loo-eet*, *yah-yah-yah*, *piiiyay* or, my favourite, *catarr, catarr* (poor pin-tailed sandgrouse with early-morning mucal problems). But it is their singing that strikes a chord in us, and it is hard to imagine that a blackbird in full voice on a late spring evening is not pouring out its soul as Rodolfo does to Mimi or Tristan to Isolde. Hard too not to think they must *enjoy* this, just as we imagine their joy in flight. In general, birds are joyful creatures that raise human spirits with their singing and swooping, and if this is not strictly speaking communication it is for me more rewarding than many a dinner-table conversation.

Birds are talented mimics, and many birds copy each other's song and calls; of our garden birds, starlings are the most gifted, mimicking domestic sounds such as mobile ringtones and spray cans being shaken. Others use mimicry to communicate with humans, and captive birds like parrots, budgerigars and mynah birds can have impressive vocabularies of several hundred words – though the difference between mimicry and genuine communication is sometimes hard to define and may be fudged by enthusiastic owners. But if you are sceptical, take a look at some of the conversations between birds and their keepers available on YouTube. There seems little question about the cognitive ability of these birds, which don't just answer questions 'parrot-like' but initiate topics of conversation and are able to express emotions such as affection and irritation. Most of them are also irresistibly perky and characterful.

My own relationship with birds is complex. I love, admire and often envy them, but I have an innate revulsion towards handling them; the spongy insubstantiality of all those feathers enveloping such a little, brittle skeleton produces a shudder. I have learned to overcome it and can now happily tuck a chicken under my arm – I even recently picked up a tawny owl that was sitting, dazed, on a cushion in my sitting room having fallen down the chimney – but to my shame the reluctance to touch them persists. However, that birds are messengers, whether from the gods or not, I truly believe.

When I lost two dogs – lost in that they disappeared during our usual daily walk and were never seen again – I went, after the initial frantic, fruitless searching, into the accepted stages of mourning. Somewhere in between the anger and depression, I began to notice the unusual appearance of birds, sometimes singly, sometimes a pair, never more. One day a robin, bold as robins are but never were before on this high land, north-west of Seville; two owls calling to each other in the valley, a first there too; a buzzard on a fencepost, staying put until we made eye contact; two azure-winged magpies that swooped low over my car, never previously sighted nearer than the Portuguese border more than fifty kilometres away. By nature sceptical, not given to mysticism or flights of imagination, I put these appearances down to coincidence, perhaps heightened sensitivity on my part in a time of distress. But it was a pair of blue tits that made me take it all more seriously.

They arrived one morning in my bedroom, cavorting around the room for several minutes without any sign of panic before flying out to play on the vine, then suddenly swooping off, weaving in and out of each other's flight-path as they hurtled down and away over the valley. The next day they came again, outside the kitchen window, alighting on two long stems of wild garlic that stuck out from the wall. They bounced up and down, one on each, then swapping, then two on one stem, then back to their own, over and over as if for pure joy, like kids on a trampoline. I watched for maybe five minutes as they frolicked – there is no other word for it – and for the first time in weeks I found myself laughing. Their visits continued for several days and each time I smiled and felt strangely comforted by their presence, as if they were telling me that my beloved dogs were indeed dead but wanted me to know that they were together, and at peace. Fanciful stuff, no doubt, but the sequel convinced me of its reality. For a few months later, long after the blue tits had stopped visiting, I decided to record their story. And as I sat writing in the bedroom where they had often entertained me with their capers, they came again, flew around for a while and then disappeared, this time for ever.

* * *

The main hindrance to true meeting within both the human and animal worlds, and between the two, is fear; whether this is instinctive or rational makes little difference to the outcome. Instinctive fear, of the adrenaline-pumping kind, sets off the fight or flight reflex – or less often, the freeze – and is therefore crucial for survival, both in the pack and for the individual, in the farmyard or the jungle. Or indeed in the living-room, for on a less life-threatening level, fear sets up many barriers and causes the nerves to jangle in human interactions just as it does in those between man and beast. Instinctive fear is immediate, urgent and untrammelled by thought processes – these may come later, in humans if not in animals. Since it is to do with self-preservation, this so-called animal fear is often seen as more 'pure' than neurotic fear. But Temple Grandin, who knows about pain in herself and the animals whose welfare she champions, asserts that fear entails more suffering for an animal, both prey and predator, than physical pain.

Certainly fear is communicated from people to animals, and the other way round. A cornered wild animal is fierce (and liable to attack) because it is afraid, and we are right to catch its fear for the sake of our own skin. But irrational fear, such as mine for handling birds or being around bees, see below, needs examining. Humans are prone to what psychologists call negative fear, the nagging anxiety that besets modern humanity and may lead to neurosis. This type of fear makes one jumpy, nervous and liable to make mistakes; it is negative in every sense, and contagious. As we will see, someone who is fearfully handling a horse, whatever the cause of that fear, will pass over the emotion to the detriment of the relationship on both sides.

Animals too suffer from what may appear to be irrational fear. I watch with amusement as my ducks, faced with their food container having been moved a metre from its usual place, approach it warily, circle round it, consult each other as to what's going on; it may take a day for them to recover their composure, and the same thing will happen if the food is moved back to its original position. These adult ducks have no predators, nothing material to be afraid of through direct experience; their fear is of the unknown, a legacy of their ancestry as prey animals. Such fear lies at the root of most cruelty on the human side and aggression on that of the animal.

True communication between sentient beings implies a state of equality, a meeting on level ground that is not possible where there is submission on one side. This is where, in my view, animal trainers may go astray. In their desire to instil obedience at all costs, the mutual understanding, and therefore respect, is lost and the animal's natural energy – however perverse it may seem at times – is subdued. It's the same narrow borderline considered earlier, between making fun of an animal and having fun with it. Only if respect is present on both sides will the meeting take place.

* * *

Whether insects are sentient is not yet proven, but that they communicate between themselves is irrefutable. Ants' communication takes the form of touch and body language, sound and scent (the latter through pheromones). They communicate among themselves to become 'one animal', as many as a million ants acting as if with one mind. Bees too are highly accomplished communicators among their own kind. The honey bees' waggle dance, which directs other bees to a source of nectar, is a sophisticated double-pronged manoeuvre that involves the positioning of the body to point the direction from the hive to the nectar supply, with the sun as the third variable, and the duration of the waggle that shows its distance. Some bees add to this already remarkable achievement with audible signals that are believed to convey the quality and quantity of nectar available.

Bees are the focus of another communication, more a collaboration, that has evolved in sub-Saharan Africa between humans and small birds called honeyguides. These have long been known – and named – for their feat of discovering bees' nests and then calling to the people, flitting from tree to tree to lead them to the source of wax and honey that each seeks. But for the Yao people of Mozambique it is they who do the summoning, having developed a specific call (and no other will do) to alert the birds for a honey

Mutual trust, mutual aid:
Orlando Yassene, a Yao, with his faithful honeyguide.

hunt. In both cases this is a mutually beneficial arrangement, as the birds
eat the wax exposed after the nest is opened – something they can't do on
their own – and the honey is taken by the people. In an account of a similar
harvest, this time in the Kalahari, the hunt and the honey are further shared
with a badger-like creature, the ratel, or honey badger. This tough-skinned,
ferocious omnivore – not a true badger – has a sweet tooth and a cunning
strategy for indulging it. Summoned by the calling of the honeyguides, the
ratel joins the hunt and on arriving at the hive, most often in an abandoned
termites' nest, places its behind against the entrance and gasses the bees;
its gas must be as powerful as the creature itself, for it knocks the bees out
and the San harvesters are able to take their fill of honey unmolested. They
reward the ratel in a solemn ritual that includes the words, 'Take, and eat',
leaving sufficient honey for the bees and wax for the honeyguides, thus
honouring 'the proportions implicit in the claims of bee as well as man,
bird and beast', a fourfold benefit.

Since bees have a sense of smell that is about a hundred times more
powerful than that of humans, with receptors on their mouths, antennae
and the tips of their legs, it is not surprising that they pick up the human
fear pheromone that even we respond to in each other.

A personal anecdote. Last summer my sitting-room became invaded by bees, a constant succession that came down from behind the wood-burner, dropped to the floor, buzzed around groggily and then congregated at the windows. I was wary, but in my old age I wanted to conquer the fear that still lingered from an earlier experience, a tragi-comical scene that I have relived many times.

It happened, with the surreal quality of a dream, like this. I was taking to my mother's unmarked grave a piece of inscribed granite that, although very heavy, I was just able to carry the short distance from car to graveside. Clasping the stone and straddling the grave as I worked out where I should place it – knowing that it would be hard to move if I got that wrong – I bent over and carefully let the stone drop. As it fell, it seemed as if in slow motion, I saw a hole in the grass just to the side of where the stone came to rest (RIP), and out of it in a flash came a horde of bees. Maybe it was my state of mind, or my instinctive fear, or just the rude invasion of their space, but they attacked me – and I panicked. I ran round the cemetery pursued by bees, whether excited by the pheromones I was emitting or egged on by their own 'attack' pheromones, I didn't know or particularly care. All I do remember was thinking wryly – for my mother and I were rarely at peace –'O death, where is thy sting? O grave, where is thy victory?'

So more than twenty years later, faced with a sitting-room full of bees – they first appeared in May and are still arriving as I write, in early November – I decided to put myself to a test. If I treat you kindly, I bargained, and don't call anyone to destroy your hive (in France this is done by the fire brigade), let all survivors out of the window many times a day, avoid stepping on you if I possibly can, above all give out friendly vibes, will you not sting me? Well, I've been stung twice in the six months I have shared my hearth with the bees, and both times they had got caught up in my clothing, which is hardly their fault – or mine. Maybe it's because I haven't directly threatened their hive, maybe they are drowsy and about to die, though many with my help have flown off into the sun or rain. Again, I don't know. But I do know that my handling of the bees, my attempt at communication rather than fear, my genuine compassion for their plight and hope for their survival has been good for me as well as adding a few more bees to the world, so maybe that is conclusion enough, however unscientific.

5
TRUSTING
ANIMALS

In draft, this chapter was headed 'Taming/Training' as I struggled to differentiate between the two concepts and find a more comfortable solution for our need to make animals fit into our way of living. The word 'tame' may be defined as 'not wild', and in its pejorative sense this implies lack of fire, lack of balls. Which is sadly often the result of taming, though its other definition, also a negative, is 'not dangerous or frightened of people'. In other words, trusting.

Training undertaken without trust – on both sides – is a mere bending of the animal's will to the demands of its trainer. It may not involve physical cruelty, but reward and punishment, whether aggressive or by denial of the reward, are usually an accepted part of the procedure, which is likely to ignore or deliberately overrule the feelings, even the instincts, of the animal. Fortunately the booted and moustachioed trainer wielding a whip over a circus lion is no longer widely tolerated and the need for kindness in animal training is recognised, but 'inhuman' methods of training, ruled by mutual fear rather than trust, still abound.

Taming and training can be effected quite quickly; they may form the basis for a deeper understanding, but solid trust is built only over time.

Developing trust is a more complex and altogether more satisfactory process than mere training, demanding patience and goodwill on both sides. Trust will not flourish in an atmosphere of anxiety or stress, and without it there is no scope for spontaneity and individuality in the relationship between people and animals.

* * *

Some training is necessary for the young of both animals and humans, of course, but the aim should be loving respect rather than rigid control. The prerequisite for an inspired and inspiring teacher, of both animals and humans, is the ability to *see* his or her charges for what they are, as individuals with differing needs and potential. I like it that the root of the word 'education' is 'leading out', a gentle process – visualise it – far removed from cramming or force-feeding or learning by rote.

Young dogs, for example, need to be trained. Not trained as in the specialised disciplines of obedience trials or tracking the lost or leading the blind, but taught what is acceptable behaviour, what is allowed and what isn't. This naturally depends on the expectations and lifestyle of each owner, and therefore varies hugely. But I believe that a dog that is trained, kindly but firmly, to have basic manners and be aware of boundaries will be happier than one that is indulged and allowed to rampage – if only because of the reactions of other people. Uncouth dogs are as hard to like as uncouth children.

In any training there will inevitably be mistakes on both sides, with setbacks and occasional dramas, misunderstandings and exasperation. It's not straightforward, and there are many dilemmas as our desire for 'civilised' behaviour clashes with basic animal instincts. For instance, my recently acquired puppy spends countless hours with her front-half down a hole looking for voles or moles or whatever it is that smells so irresistible to her. She digs furiously, feet pummelling, shoulders braced, teeth tearing at turf and roots, every ounce of energy focused on the task. This is clearly good for her, developing her sense of smell, engaging her attention, strengthening her muscles. I enjoy watching this total absorption, seeing her instincts at work, *seeing* her, and it's fine for me too so long as it takes place in the woods or fields. But in the flower bed or on the lawn it is not so popular, and I have to teach her this boundary that doubtless seems incomprehensible to the dog mind. It's a question of letting out the rope and then hauling it in, over and over, so that the trust builds. It should be a journey not an assault course, for an animal can be forced into obedience but if there is no trust the relationship will be robotic and will not evolve into one of mutual gain. A dog, particularly, that is trained through fear will fail to develop its personality; like a child who has lacked love and understanding in early years, it will bear the scars of this throughout its life. My dog Danny, a stray who had probably been abandoned – he was distressingly hungry at the time he was found, but not malnourished – came to me at about a year old. After eight years I would say that he trusts me, but if I pick up a stick clumsily he still cowers, and his love-bucket has a hole that will never be plugged.

I have come to believe that animals can play an important part in training their owners too. As one who myself lacked that early love, I went about looking after the first dogs in my life heavily weighted with my parents' rigid ideas about routine, structure, discipline – not one of these in itself harmful, of course, but they come with hard edges that need the lubricant of love if they are to bring out the best in both the carer and the cared for. Now, many decades later, after much heartache and soul-searching (as well as a lot of fun) and six more dogs, I think I might be getting the hang of what they instinctively know about patience, endurance, forgiveness and trust. Above all they have taught me about loving – real love, not the mushy sort. And there's plenty more to learn.

* * *

Some years ago I was elected – by him – as the guardian of a stray dog, a mastin (Spanish mastiff) with a large shot of husky blood, whose route to my doorstep makes too long a story to relate here but who was known locally and was guessed to be about six years old when I took him in. The first part of Thulo's life had been as a vagabond, until he became attached to a young man who himself lived a peripatetic and unpredictable life, so this was a dog not used to routine or containment. Thus began my true education in dog psychology, and a steep learning curve in patience.

There were two, or perhaps three, things in favour of its working. First, I myself have a phobia about containment: a man in prison, a big cat in a cage, a small cat in a high-rise flat, all put me into a frenzy of anxiety and fellow feeling. Second, fortunately for Thulo, we were living in rural Andalusia where space was not at a premium and life was not yet rule-bound, and provided that your dog didn't chase livestock or bite children, no one much cared what it did. The final thing going for this dog was his laid-back, good-natured personality. Everyone loves him and he loves life – so long as he is free.

At first, when it became clear that he was here to stay, I misguidedly tried to 'train' him. I closed the gate, put him on a lead (*a lead?*) to take him for a walk (*do you call that a walk?*), tried to instil a routine (*what now, when I'm sleeping?*). I prepared a bed in a shed that he scorned, preferring to sleep outside in sun, rain, frost or even snow – the husky genes gave him beneath his glossy coat a winter under-blanket of dense wool that made him impervious to cold, and he knew how to seek out refuge from the Spanish midday sun. He never came inside the house, regarding the tiled floors as ice-rinks that were not at all to be trusted.

But between us the trust grew. Very slowly, as is the way with animals, we began to know and to make way for each other. I no longer shut the gate, for it made no difference; if he wanted to go somewhere, he went: through thorn hedges, down perpendicular banks, over walls, *through* walls it seemed; nothing stopped him. When I went for a walk he would shamble up the road with me for a while, then peel off and go his own route, where the smells were better and there was the chance of some pickings (it took a long while for him to expect food to arrive on a regular basis).

In the eight years since he came to me – I would never claim to own him, and through him would not now say that I owned any animal – we have evolved a *modus vivendi* that has left us both more than satisfied, and myself immeasurably enriched. Thulo doesn't 'do what he is told', and never will, but he knows what is approved of and what is not, and he shows his affection and gratitude in unmistakable ways. He rocks up, stiff-legged, to greet me first thing in the morning, pushes his great weight against my legs, rolls over for a tummy rub. He has learned to play, with me and with Danny, another mastin cross who arrived a year after Thulo and much to my surprise was welcomed by him. Both dogs, who had never before left their home territory, travelled by car from southern Spain to northern France, where they now enjoy a more temperate climate and even greater freedom. Thulo is old and lame, plagued by long-ago injuries and arthritis, but still takes off at impressive speed if he senses anything worth a chase and is heedless of calls to desist, though looks suitably penitent when admonished afterwards for chasing the post van. He is the undisputed top dog, and male canine visitors are not tolerated.

Thulo,
aged and arthritic
but never other
than alpha male.

It is hard, impossible, to convey the character of this dog, who looks into your eyes with total trust yet without any suggestion of grovel. He is dignified, patient and wise – anthropomorphic adjectives that as well as embarrassing me, themselves sound lame. Names given to him by his fans include King Dog and Great Bear, for Thulo (the word means 'large' in Nepali) is a big dog in both body and soul. He has a quality of gravitas that is in no way compromised by his habit of rolling on his back with all four paws in the air. He doesn't fit into any dog-plus-keeper category that I have heard of, for he is entirely autonomous. He still doesn't come into the house and hasn't learned to ask for anything; if he's not around at the normal feeding time and I forget his dinner, he doesn't remind me. He has no *need* of human company while enjoying it hugely when it is on offer. Our communication is almost entirely non-verbal, but we have learned what to expect from each other and are comfortable with that. We trust each other. And I love him. When I go out late on a rainy night to the barn where he now, in his old age, condescends to sleep if the weather is bad, it is because saying goodnight to him gives me the shot of oxytocin that will send me happily to my own bed.

* * *

Let's stay with dogs, the first animals ever to be tamed and trained by man and still the ablest communicators of all. On a long journey from their first use as herders, dogs have now become helpers and companions not only to the blind but to people with many different disabilities, physical and mental. Imagine how this could transform your life, for as well as being trained to do an impressive range of fetching, carrying and guiding tasks, these dogs offer unconditional love and loyalty to people who might otherwise feel helpless and abandoned. The mutual caring is a mutual gain, and the dogs seem to respond to their role in an extraordinarily sensitive way, able to pick up on the smallest sign of distress or need. There are dozens of affecting stories online about assistance dogs, especially their ability to communicate with autistic children and with people suffering from dementia. It seems that an animal, and in particular a dog, can short-circuit the brain and communicate directly with someone who otherwise would be locked up inside a prison of the mind. A nine-year-old autistic boy, Toby, whose behaviour was transformed when a black Labrador was introduced into the house, puts it most beautifully: 'We're connected by an imaginary string from his heart to mine.'

Animals open hearts. Disturbed by the atmosphere of boredom and inertia in the nursing home that he had joined as medical director, a young American physician named Bill Thomas decided to shake things up by introducing animals into the sterile environment. Battling his way through the inevitable protests of the health department as well as some of the staff, Thomas got his way and amidst 'total pandemonium' two dogs, four cats and a hundred parakeets were installed on the same day, along with live plants (replacing the plastic ones) in every room. The results were startling. The patients pitched in with feeding and caring, and those 'who we had believed weren't able to speak started speaking. People who had been completely withdrawn and non-ambulatory started coming to the nurses' station and saying, "I'll take the dog for a walk." All the parakeets were adopted and named by the residents. The lights turned back on in people's eyes.' What is more, a two-year study showed that the amount of drugs needed by patients fell to half that of a nearby nursing home used as a control, and deaths were lower by 15 per cent.

Although their desire to communicate with humans gives dogs the edge as helpers, many other animals now play their part. Horses and donkeys have

Determination and dexterity:
a capuchin monkey
with its distinctive cowl.

proved themselves trustworthy therapeutic companions without specialist training, and some birds, parrots in particular, are highly responsive to mood in humans, able to give warning of incipient stress disorders such as epileptic fits and bipolar attacks. Monkeys too, the tiny capuchin especially popular because of its size and dexterity, are ingenious helpers who seem to enjoy the tasks set them. It may sound fanciful, but these animals appear to *want* to help people. Maybe this is in direct response to the trust invested in them.

* * *

Training a horse for riding is a different matter, and it is significant that had this book been written even twenty years ago, I would probably have put 'breaking' a horse. The word speaks for itself, and horses still routinely get broken in many parts of the world, as draught animals, broncos and hacks. (Other animals – people too – get broken; you can see it in their eyes. It happens through human cruelty, not natural adversity.)

Humans have been training horses for a very long time. There's a text on training for chariot racing from as far back as around 1,400 BC, its focus on how to bring the horse to peak condition. The first known treatise on horsemanship, written in about 350 BC, is by Xenophon, who famously regarded it as an art but whose advice – mainly aimed at training and handling horses for battle – might not chime with today's horse whisperers. The work begun by Monty Roberts in the 1980s and adopted by many others since has changed attitudes and continues to spread – and cause controversy. For it is easier, and quicker, and requires less skill, to 'break' a horse, even though doing so may also break its spirit and affect behaviour and performance for the rest of its life. But short-term gain is alluring for those whose main interest is making a quick buck rather than training a horse for life, and the turnover in horses can be rapid, a point harrowingly made in the tale of Black Beauty.

Vicki Hearne, already mentioned for her ability to communicate with horses at a deep level, approaches the training of them – particularly what she calls crazy horses, messed up by someone else's attempts at training – with the questioning mind of a philosopher and a dogged determination to understand everything there is to be understood about each one. Hers is a voyage of discovery in which the horse participates fully; Hearne uses her formidable intelligence to 'read' the animal – to *see* it – and then to shamelessly outwit it. There is nothing sentimental about this; she herself labels sentimentality an infection, and her approach is tough and her standards high. This is horsemanship – or, if you must, horsewomanship – of the highest order. Few are so dedicated, and training is for the most part a haphazard affair resulting in the horse being 'good enough' – good enough to be ridden, hacked, played around with and often left to its own devices for long periods.

A prerequisite for successfully training a horse, at whatever level and for whatever purpose, is mutual confidence. Being a prey animal, the horse needs to feel comfortable, and for this the person undertaking the training needs to be similarly at ease. As we have already heard, horses will pick up on an angry face, and angry vibes or movements will disturb it more deeply. But it is more subtle than this. Many people maintain that horses discern what is fake, and that it is impossible to hide negative feelings even if they are accompanied by seemingly gentle actions. One trainer goes so far as to say that anything motivated by the ego is liable to be thwarted when working with horses, and many claim that they play a part in raising human self-awareness. Training horses is certainly a learning experience on both sides, for those willing to learn. Alois Podhajsky, director of the Spanish Riding School in Vienna and a renowned teacher of classical equestrianism, named the highly acclaimed book about his life's work *My Horses, My Teachers*.

* * *

Some may think that mutual trust is an irrelevance in our dealings with any animal, but except in cases of blatant exploitation, most people who live with animals, who know and *see* them whether in the home or working with them on the land, in zoos and sanctuaries or in the wild, will strive for some degree of trust in their relationship. Trust is the foundation upon which other things may be built. We don't know how long it took for the taming and training of dogs after the cave dwellers first threw scraps to the wolf-dogs and realised the possibility of a mutually beneficial arrangement between them, but it has taken millennia for trust to be fed into the genes of both dogs and humans. And each time the individual dog and its keeper must start again their own building process, a unique journey with no rules, no referee, no gain except that of wellbeing and satisfaction on both sides.

Nor is this confined to the animals that share our hearth. The wellbeing of those in the stable and farmyard will affect the work done (once with cart or plough, now perhaps in the show ring), the milk yielded, the meat and eggs produced. For stress in any form has a negative impact on the animal psyche, which flourishes in an atmosphere of trust and calm. If this sounds far-fetched, count the eggs after a drama has occurred in the hen house, see the milk yield drop when an angry helper is introduced to the cow parlour, taste the difference in the meat of pigs that have snuffled and wallowed in open ground and those having spent their lives in a concrete barn. It may be stretching it to attribute this to mutual trust, but based on my own observations I hold that if animals are kept in a natural environment and treated with reverence, the rewards will be more than just an untroubled soul on the part of the keeper. Caring custody breeds trust, and trust is as contagious as is its enemy, fear; both spread their ripples wide.

The gradual process of taming and training domestic animals, whether resulting in trust or not, has had an immeasurable bearing upon our own evolution. Those wolf-dogs slowly – for evolution is immeasurably slow –

became herders, which made possible containment, and from that came selective breeding. People with large herds were less inclined to move on as nomads, so they became farmers; then they needed help on the land so they turned to the beasts and harnessed their energy. These developments took place over many thousands of years, but without them the human race would not be where it is today; we needed the help of animals while we were slowly gestating machinery, then more rapidly technology. And we need them now, to keep us in touch with Nature, and with our own true nature, in the stifling air of our increasingly urban lives.

There are no rules in the strange relationship between human and animal, whether parrot, pony or porpoise, and many variables that will affect the level of trust achieved: the inherent character on both sides, the cultural backdrop, the needs and expectations of the human and the ability and willingness of the animal to respond to these. As with human relationships, they may evolve or get stuck, be fruitful or sterile, bring joy or misery. But we as humans, with our huge advantage of brain power, hold the key to the outcome of each unique journey undertaken by people and animals, through taming and training, to the ultimate reward of mutual trust.

6
HUNTING
ANIMALS

We were all hunters once; it is how we survived, how we ultimately, after a couple of million years of evolution, separated ourselves off from others in the genus *Homo* by developing superior weapons and methods of trapping and killing prey. We humans are omnivores, our teeth designed to eat meat as well as fruit and vegetables, although we can – and maybe now should – happily survive as vegetarians.

Hunting is in our blood. As a child I was brought up to regard hunting – that is people on horseback following a pack of hounds following a fox – as normal and admirable, and I have a vivid memory of being 'blooded', arriving at the kill and having the fox's hot blood smeared over my forehead. I didn't find it disgusting then, although it seems so now. In my late teens I followed a university pack of beagles, though my attention was focused more on the male students than the hare – hares, for heaven's sake, those bewitched and bewitching animals that I would now do all in my power to protect. Attitudes change, both for society and for us as ageing individuals; children are more casually cruel as they explore their world, and the small boy who delights in pulling the wings off flies may end up as a dedicated entomologist. But old age brings

Big cats and small, whether in the wild or the back garden, share the same instincts; but the survival of this very young mountain lion cub (top), which still has much to learn, will depend on its success as a hunter, whereas for the domestic cat there is no such need.

empathy, as our bodies become frail and blood shed in wanton killing shocks us the more.

So the term 'blood sports' must be an oxymoron: how can we kill for fun? Yet we have done so for thousands of years, ever since the hunter-gatherers turned to agriculture and hunting for survival became unnecessary, and we continue to do so today when the camera allows us to shoot animals without spilling their blood. It seems to be in our genes, in the male genes particularly, a throwback to the time when the man brought home the bacon and the woman cooked it. In the leisurely pace of evolution it will take many more millennia before we lose the desire to kill animals that we do not need to eat.

For the hunting animal, there is no such thing as cruelty: it must kill or it will die. In this context, 'cruelty' is a concept invented by humankind. Wildlife programmes bring into our sitting-rooms stark images of wolves working in packs to bring down the weakest, most vulnerable moose, of bears gobbling fish, crocodiles crunching antelope, leopard seals grabbing penguins. This may seem remote from our own lives, scenes from a wilderness that does not impinge on us, but you have only to go out into the garden to see much worse. In the behaviour of the domestic cat there is something that does appear to be gratuitous cruelty as it stalks, pounces and then spends the next half-hour tormenting its prey, only to tire of the kill and stroll into the kitchen to demand its regular supper of Kitekat. Thankfully the cat seems to be unusual, if not unique, in its sadistic tendencies, but throughout the animal kingdom instinct remains long after need has been removed.

* * *

Human beings too are subject to strong instincts, which are worth considering in the timescale of our existence on this planet. Our hominid ancestors are thought to have emerged around 2.8 million years ago, but *Homo sapiens* not until 200,000 years from the present. The next big step in our development, our becoming farmers, took nearly 188,000 years, during which time we survived by hunting and gathering. This makes our subsequent 12,000 years as farmers seem rather paltry – but not as paltry as the 200 years during which we have become largely urbanised. So it is not surprising that we retain many of the impulses that date back to our time as hunters, by far the largest slice of human history.

Our minds are shaped by our lives as hunter-gatherers, and it seems that we still, if only subconsciously, belong to and long for our 'pre-civilised' existence. This may be no more than nostalgia for a fictional Garden of Eden, but the longing persists. Our instinct for gathering, even if no further than from the vegetable patch, is the more acceptable side of this often romanticised state, but the desire to hunt is just as strong – and throws up some interesting speculation about the gender roles in these two activities. It's a controversial topic and this is not the place to go into it in any depth, but it is undeniable that few women go shooting for pleasure. They may be just as good at it, as skilful at aiming and as capable of wielding a gun, but most of them just don't want to do it. It was not women's role as hunter-gatherers; they did the gathering (though men did that too) and the cooking, but not usually the hunting or the butchering, which were men's work. It was partly a matter of physique, partly the necessity for teamwork in raising children, and for most of us these instincts remain strong.

Meanwhile, the men go on hunting. It's a spectacular sport, the sport of kings (a label also given to horse racing), with all the attached hierarchical display. Uccello captured it beautifully, and without any bloodshed, in

The Hunt in the Forest, a feast for the eye in composition, colour and perspective, a romantic version of what is in reality a muddy, muddled and brutal activity. Here the aristocracy are mounted; the rest, hunt servants as they were and are still called, went on foot and tried to ensure that the prey – in this case deer – went in the right direction. English monarchs have always been keen hunters, despite the death in the New Forest of both Prince Richard and William II, the latter by an arrow from an unknown hand: the hunter and the hunted.

There is a lot of pageantry in the hunt. People have dressed up to go hunting ever since it became a sport, and often the dress serves no other purpose than mere show. Huntsmen may have originally worn scarlet coats so that they could easily be seen and distinguished from the rest of the human pack, but this didn't help the hounds – nor does it explain why this scarlet is called pink. But the tradition remains, just as French shooters today still wear camouflage jackets and trousers under the day-glo vests they have adopted, not to alert their prey but to save more of them from being shot by their fellow hunters after a good lunch.

There is also much snobbery associated with hunting – indeed with horse-riding in general, as the horse physically elevates the rider above the common crowd. Even when horses were the normal means of transport, the breeding of your horse mattered in social terms, and the poor were equally competitive about the donkeys and mules they rode when they couldn't aspire to owning a horse. Pride of ownership continues to this day, as people show off the latest model of car to signal their one-upmanship. (It doesn't always work. An anecdote that I prize from my time in Andalusia tells of a nouveau-riche local man who developed a taste for flashy cars. Parading through the village in his latest acquisition, a Rolls-Royce, he passed two old men sitting on a doorstep. 'Ah,' said one, 'so he's changed his donkey again.')

Foxhunting became increasingly popular in Britain after the enclosures that began in the mid-sixteenth century and reached their peak at the end of the eighteenth, bringing misery to both humans and the animals such as deer and wild boar that had previously wandered freely over large areas of forest and common land. The royalty and gentry who had then hunted these animals – and preserved the forest to do so – now turned to foxes for their sport, taking in their stride the hedges and ditches that enclosed the fields but were no hindrance to the fox. Twenty-odd years ago I watched a foxhunt in full cry crossing an agricultural plain in Hertfordshire – few

Hunting for necessity.
Cave painting from Tassili n'Ajjer in the Sahara, c. 7,000 BC.

Hunting for pleasure. Note the double fence to make it more 'sporting'.
A scene by Henry Thomas Alken (1785-1851).

hedges remained – with the white of the hounds and the huntsmen's pink picked out by a shaft of sunlight piercing a plum-blue sky. It would be hard not to react to such a scene, whether in outrage at its barbarity or appreciation of its beauty – mine an uncomfortable mix of both.

I have no sympathy for abolitionists, whose often violent behaviour seems only a crazed mirror to what it is they are protesting against, but the dividing line between squeamishness and sentimentality on the one side and unmerited cruelty on the other is a difficult one to negotiate. Often it signifies the split between town and country dwellers, for acceptance of death comes with close contact with Nature. The lives of animals are shorter than our own, their ending often sudden and violent. The natural order provides for this 'wastage', and most animals reproduce in far greater numbers than we do, the loss of one providing the means of survival for another.

We humans have persistently interfered with this order, upsetting the fine balance of Nature and creating vacuums; whether the motives were of exploitation or necessity is irrelevant to the outcome. Some time after seeing the foxhunting tableau described above, I watched rabbits hopping about the grassland nearby in numbers that seemed obscene, like fleas on an old dog. Fewer foxes equals more rabbits, and killing them with myxomatosis was a solution that not many could stomach.

George Monbiot supports the rewilding of certain designated areas of land across Europe and would like to see wolves reintroduced to Britain. As back-up to what some – predominantly landowners – consider this irresponsible idea, he cites the return of wolves to Yellowstone National Park, a massive area of almost 9,000 square kilometres in Wyoming. In a misguided attempt to protect the numbers of red deer, wolves there had been deliberately exterminated in the early 1900s, and thereafter the whole habitat had changed rapidly. Monbiot describes what is known as the trophic cascade, the effect of the removal or severe reduction of predators at the top of the food chain that tumbles downwards, affecting not just the animals lower down the chain but the entire ecosystem – trees, vegetation, the composition of the soil and ultimately even the climate. Hunting in the animal world has a purpose beyond the survival of the individual or even of the species; it plays an essential role in the delicate, intricate balance between animals and the environment, and we interfere with this to our peril – and theirs.

* * *

It is ironic that hunting was more of a 'sport', that is more a matter of chance and less loaded in favour of the hunter, in the time of the hunter-gatherers than it is now with our breeding programmes and sophisticated weaponry. But the animal spectator sports are a different story. They can't really be classed as hunting, of course, but they spring from many of the same instincts, and it is perhaps encouraging that the most extreme types such as bear-baiting and cock-fighting are now seen as barbaric and in most countries are illegal, though they persist underground. More controversial is bullfighting, still considered an art form by many people, and surrounded by so much tradition that its loss would leave a gaping hole in Spanish culture. In *On Bullfighting*, A.L. Kennedy writes with feeling as she explores the masculine world of the bullfight, facing up to all the usual objections as well as adding some of her own. No one should make up their minds about this extraordinary ritual, in which communication between man and animal undoubtedly takes place, before reading this book. Is bullfighting an art? Yes, though one may not go so far as Lorca in describing it as a religious mystery. Can it therefore be justified? No, of course not – but nor can bombing citizens in Syria. The capacity for maiming both fellow men and animals, whatever the motive, is endemic to the human race, and although as individuals we may attempt to grope our way through the moral maze and make our own decisions regarding what is acceptable and what isn't, it is indisputable that violence is in our history and our genes, and it is likely to remain. 'There is a wolf in me. . . . I sing and kill and work. . . . I came from the wilderness,' wrote Carl Sandburg.

* * *

Sport is often seen as a 'safe' way of keeping in contact with our primitive nature, which might otherwise come out in more destructive forms, or turn inwards into neurosis. Ted Hughes makes this point when defending his passion for fishing. 'After sixty years of experience it seems to me that rod and line fishing in fairly wild places is a perfect hold-all substitute for every other kind of aberrant primitive impulse.' Hughes knew about fish, as he knew about most animals, and few who have read his poems could accuse him of lack of feeling for them. He dreamed about the fish he caught and they came to him in poems, as did the fox and the crow. James Hillman, whose life study was psychology, wrote: 'Let the animal teach you. For the dream animal shows us that the imagination has jaws and paws, that it can awake us in the night with panic and terror or move us to tears.' We may hunt them or fish them or dream about them, but if we listen to the animals – if we *see* them – they have much to teach us.

Falconry (the word includes hunting with all birds of prey) is another solitary, controversial and ancient sport, described in the 4,000-year-old *Epic of Gilgamesh*. Falconry has its own language and rules of conduct; it is esoteric, and to me as a teenager it appeared unashamedly romantic, a masculine pastime summoning up hawk-nosed, sunbeaten faces lifted towards huge birds on gloved fists against a backdrop of sand dunes, along with the Arab horses of my dreams. This image was replaced by something more down to earth when the film *Kes* came out, and finally shattered for ever by a book, *H is for Hawk*, the account of Helen Macdonald's struggle to come to terms with her goshawk, her father's death and herself. Flying a bird of prey is a one-to-one occupation that demands mutual trust and close cooperation between bird and human – at some cost to the latter. It is no light undertaking, involving dedication, discipline and anxiety in bucketloads. T.H. White, another falconer whose story runs through Helen Macdonald's, suffered most terribly during the training of his own goshawk, which he described as a spiritual contest – and one that he eventually lost, along with the bird.

So why do people choose to put themselves to this harsh test? There's a lot of therapy involved, in all recorded cases (though maybe only those looking for answers to their own problems are moved to record them). For Billy Casper, the boy hero of *Kes*, training the kestrel lifts him out of the miserable family life and bullying that have been his lot, building his confidence and self-esteem. The film ends tragically with the death of Kes, but we are left feeling that healing has been achieved. For T.H. White there was no such resolution. Equally wounded by his very different social background, White had a vision of becoming as free, as feral, as the hawk he attempted to tame. But he ended up re-enacting with the bird his own damage, finally seeing himself as the tormenter. Macdonald ponders this anguished journey as she passes through her own rites of passage with her goshawk. More robust than White, perhaps more responsive to the reality of the bird's needs and limitations, through

Billy and Kes.

it she faces her own demons and can move on, secure in the 'protective spirit' of the bird who has seen her through. Animal therapy can get no more intense than this.

* * *

A strange word 'shoot', spanning as it does the opposites of destruction and new growth. Photography lessens the divide, and this latest form of hunting does perhaps hold out hope for the future preservation of our animals. Partly through political correctness, partly public opinion, but in part too to a genuine concern over the fragility of many animal species, people who once would have gone on safari armed with a gun now take instead their camera. The tracking, if mostly from the safety of a four-wheel drive, is a similar process to that employed by our ancestors, and the thrill of a sighting as satisfying even if the appetite is not so urgent. Old film footage of white men in pith hats posing with one foot on top of a slain elephant seems to us now repulsive, as proven by the recent outrage against an American who shot the venerable lion Cecil in a national park in Zimbabwe. Yet poaching of wild animals continues on a huge scale, particularly those that supply the market for cosmetics and supposed aphrodisiacs, and photos that might have shown these animals in their living glory bring us instead images of abandoned corpses with the relevant prize hacked off. Our hunter forebears would surely have been as shocked as we are.

Attitudes, thankfully, change. Karen Blixen, unusually for a woman, was a renowned shot. She writes: 'When I first came out to Africa I could not live without getting a fine specimen of each single kind of African game. In my last ten years out there I did not fire a shot except in order to get meat for my Natives. It became for me an unreasonable thing, indeed in itself ugly or vulgar, for the sake of a few hours' excitement to put out a life that belonged in the great landscape and had grown up in it for ten or twenty . . . or a hundred years.'

When it is not done to feed your dependants, hunting – shooting in particular – is a supremely selfish act. As John Fowles put it, 'You don't just shoot a deer one evening; you also shoot a piece out of every other human being who might have seen and enjoyed the sight of it and its progeny, had it lived.' Fowles, a naturalist and in later life an ardent conservationist, had been taught to shoot in childhood and was explicit about the thrill of hunting, but gave it up when 'one day I found I couldn't live with the enjoyment I got one moment from seeing birds and animals in the wild and the enjoyment I got the next from killing them'.

Equally articulate, equally ambivalent is Charles Foster, whose graphic description of shooting red deer on a Scottish moor ricochets from extreme to extreme: 'I loved it. I love it still.' 'I was doing a terrible, intimate, undoable thing.' 'I felt dirty.' 'I should have concluded that killing them was more morally serious than I did. I didn't, because no one, and least of all me, is morally consistent, and because I was having too much fun.'

Excitement, enjoyment, fun; ugly, dirty, undoable. How can such words apply to the same thing? I am aware that my own ambivalence shines out of these musings. Humans are contradictory creatures, and just as the trusted pet can turn on its master, we too may fail in constancy. I hate the wanton cruelty of cats but delight in the beauty of their preying movements, deplore but am fascinated by the hunt in all its dashing pageantry, regard the bullfight as barbarous but was once riveted to my seat by one. And so on. What is certain is that hunting within the animal kingdom must continue because it is necessary for survival, and that humans' hunting of animals, despite our modern squeamishness and awareness of dwindling species being unnecessarily wiped out, is likely to continue because it is in our blood too.

7
Using
Animals

The distinction between our use and abuse of animals has become increasingly blurred as a rapidly growing human population and desire for cheap food lead us to ever more heartless ways of keeping and treating those animals – still misleadingly referred to as domestic, when their quarters are now largely industrial – which provide the meat, fish, eggs and dairy products that we need in ever-increasing quantity (never mind the quality). It's hard to see this situation improving, unless Nature revolts by introducing new diseases brought on by overcrowding and the denial of basic 'rights' for animals such as freedom of movement and fresh air.

The confining of animals, which in the far distant history of farming may have been as much to do with their own safety and welfare as with our need of them, has reached unthought of levels. And lack of thought, let alone empathy, is the predominant feature. The plight of chickens spending their entire lives in batteries, cows spending theirs on revolving production lines, pigs transported for hours, even days on end in containers – consider that word – is well recorded, and mostly accepted as reasonable. On my neighbouring dairy farm in Brittany, perhaps a couple of hundred hectares run by honest, good-hearted people who have farmed there for generations, the calves never get to gambol in the fields. The females are kept inside until well past the age of gambolling, while the male calves spend their short lives eating, sleeping and shitting in tiny metal cages, soon to become veal. (I have just read an advertisement for calf pens that lists twelve advantages for the farmer but never once mentions the calves.)

You don't have to be an animal rights activist to be horrified. Any half-caring person, if forced to sit and think about what this means, must surely react. It is inhuman – another word to consider – to prevent young animals from playing and from exploring their world, from sniffing and skipping and doing all those things that are vital for their learning and incidentally so enchanting for us to watch. Their adult life, of confinement if not containment, is dull enough for sure; cows, when not loose on the prairie, face little danger and spend their days munching and wandering around their safely fenced pasture, weighed down by the immense udders that justify their existence as milk machines. These are the lucky ones, though, for many grazers never even see grass. Preventing a grazing animal from grazing strikes at the roots of its being; it is a deprivation beyond imagining. But this is on the smaller end of the scale of misuse that we now take for granted in our desire for cheap milk and meat; it is not overt cruelty, not detectable abuse, but something more insidious, a wilful neglect of the true needs of creatures that we exploit for our own purposes.

For the paradox of any form of containment is that it keeps safe but at the price of freedom, something for modern humans to ponder not only in relation to animals: exploration and risk enliven the spirit, safety and boredom slowly suffocate it.

* * *

People themselves contained turn to animals, as symbols of freedom, as inspiration or as fellow sufferers in their own predicament. Famous for this was Robert the Bruce, King of Scotland, on the run and holed up in a cave (or a shack – there are many versions of this story), who idly watches a spider attempting and failing again and again to swing its thread and secure the web it is making. His attention caught, Robert at last *sees* the spider and draws the well-worn moral, 'If at first you don't succeed, try, try and try again.' Brian Keenan, wrestling to keep his humanity in what he calls his 'creature-condition' while in solitary confinement in Lebanon, identifies with animals in captivity: 'I thought of animals in the zoo, with their desperate patience or spirit beguiled into some neurotic state pacing to and fro, their minds empty.' Later, in a cell with two others, he becomes maddened by an invasion of ants and his first reaction is to 'stamp and slap and crush them, without mercy, without any thought of their separate existence. But after days of this I got tired of my anger.' So, like Robert the Bruce, he begins to *see* the ants. 'My fascination made friends of them. I was grateful for their fortitude, for their strength, for their resilience and instead of raging at them I would sit awaiting their return.' Watching as the ants labour to carry an injured companion to be amongst its own, he says, 'This incident became a symbol for me in this blank room with its three chained creatures. We cannot abandon the injured or the maimed, thinking to ensure our own safety and sanity. We must reclaim them, as they are part of ourselves.'

From a very different but equally horrifying prison, that of locked-in syndrome, Jean-Dominique Bauby painstakingly dictates by blinking one eyelid a book that he calls *The Diving-Bell and the Butterfly*, an image that needs no explanation. His imagination takes flight as he lies, immobile and powerless, listening 'to the butterflies that flutter inside my head. To hear them, one must be calm and pay close attention, for their wingbeats are barely audible. Loud breathing is enough to drown them out. This is astonishing: my hearing does not improve, yet I hear them better and better. I must have the ear of a butterfly.' Butterflies, symbol of the spirit breaking from its cocoon and flying to freedom, of life after death, were drawn and scratched in their many hundreds on the walls of the Majdanek concentration camp in Poland, and subsequently became the emblem for the 1.5 million children who perished in the Holocaust.

* * *

We have jumped in at the deep and dark end of the many ways in which animals serve us, physically and metaphorically, but human love for animals persists, even though it may take us down some strange paths. Our lives remain intertwined, however remotely, our involvement as strong today as our dependence on them has been throughout human history.

Take dogs, the first animals with which we developed a relationship, and that the closest of all. They were originally used, and still are, as an alarm against intruders, animal and human. For this service they were doubtless thrown the odd scrap of food, and thus began the slow story of mutual trust and mutual gain that led to selective breeding of the domestic dog, the most varied species of mammals in the world.* Dogs have certainly been used, just as they have been abused, but they have gained a lot in the process. No one who knows dogs would disagree that the pleasure between them and humans is mutual, though it is the owner or handler who can make or break this relationship. Most – but not all – problem dogs are the result of lack of love and/or discipline, one or the other being essential and a combination of both ideal.

Domestic animals have lost their genetic freedom, wrote Desmond Morris. With dogs in particular, selective breeding has stretched the possibilities to extremes, from Pekinese to poodle to Pyrenean mountain dog. But it is the extraordinary range of size, shape and temperament that has made dogs the most popular of all animal companions: there's one to suit all tastes and settings. And although breeders have played with genetics to produce some very fine and some very sad specimens, the latter at the cost of both dignity and health, dogs have a flair for escaping when that tempting little bitch down the lane is on heat, and they have defied their owners and Morris by coming up with some pretty wild mixtures themselves.

Personally I'm a fan of mongrels. I like the element of chance involved, the mix of characteristics, the unique qualities that no pedigree can bestow. But the world of specialist breeders and pedigrees is big business, and the showing arena is known for its cut and thrust as much as its cut and blow-dry. Not just for dogs, of course; cats get the same treatment, and horses too, though they are usually judged on performance as well as looks. And although the whole business of competitive breeding and showing sickens me, I have to admit to going weak at the knees at the sight of a pure-bred Arab horse showing its paces in the ring – but I would still prefer to see it in the desert, where it came from.

---

* The dog family, the *Canidae*, includes wolves, coyotes, jackals, foxes and dingoes.

* * *

Dogs have many uses, even today in a mechanised, urban world. We've already taken a look at the training of dogs as helpers for people with many kinds of disability, both physical and mental, and other animals such as horses and donkeys are being increasingly used in therapy. It's easy to see why. As well as the instinctive empathy that most people, and almost all children, have for animals, it is the non-challenging, non-judgemental nature of the relationship that works its magic.

The dogs trained for helping the blind and disabled may inevitably suffer a certain degree of containment, but there is plenty of evidence to show that they enjoy their work. Of all animals, dogs thrive the most on human attention; as a generalisation – although there are exceptions among both breeds and individuals – they seem to need the mental stimulus that being part of human life entails. If we curtail the physical freedom of the animals that we have invited into our homes, we can to some extent redress this by providing mental stimulation in compensation for that boring, repetitive walk around the park.

Boring and repetitive work is part of life for most people, whether farmers or commuters, and it is certainly the lot of working animals. Ploughing, sowing, harvesting, transporting by means of animals still goes on in the twenty-first century in a surprising number of places around the world, and for many people draught and pack animals such as horses, donkeys, mules, bullocks, oxen, water buffalo, yaks, reindeer, llamas, even elephants may make the difference between survival and starvation. (Westerners whose methods of carrying out these labours are now almost entirely mechanical, still talk about harnessing energy, and measure the engine under the bonnet in horsepower.)

Including horses among the draught animals is doing them a disservice. Although dogs were domesticated at least ten thousand years – possibly many more – before horses, the taming and subsequent use of horses had an even greater impact on the way people have conducted their lives, right up until the early twentieth century. Horses seem to have been designed for man's use. The stocky little animals we know from the cave paintings were hunted for their meat over millennia before being corralled, milked and bred. Unlike the cow and the reindeer that were first used as draught animals, horses are not ruminants and so are able to eat and move, and

Shire horses like these are probably descendants of the destriers ridden by medieval knights; they can stand up to 21 hands (over two metres) high at the withers.

therefore work, without the need to lie down frequently and chew the cud. Their physique, strong and fleet of foot and conveniently shaped to carry a rider, was another huge advantage. But most important of all was their sociability. Being herd animals with a pecking order, they were easily dominated by their early keepers, and their innate desire to please and willingness to cooperate stands out from all other animals of this size. Horses are one of God's greatest gifts to mankind.

The list of their uses is more varied than that of any other animal. Horses were, sometimes still are, used to pull ploughs and harrows, water wheels and threshing boards, sleighs and toboggans, chariots and charabancs, mail coaches, taxi cabs, hearses, state carriages and the ubiquitous humble carts. Imagine the lifestyles spanned by these uses, from the top to the bottom of the social scale, and the variety of trades required to support them. For thousands of years people were dependent on horses, in towns as in the country and on all routes between. And this is just the pulling; the carrying is every bit as colourful. Imagine these horses, *see* them as the cavalcade passes, bearing Roman emperors, medieval knights in armour, cowboys and Indians, post riders and picadors, clergymen and doctors, farmers and tradesmen. Horses were bred for these diverse roles, for strength or speed or stamina – in some cases all three – as the job demanded.

Many of these tasks were humdrum at a time when the daily round was hard for humans and animals alike. Those in power could make life miserable for the underlings, and they in turn could mistreat their charges. But drudgery doesn't have to entail suffering, and now that knowledge is more freely available and awareness of animal sentience widespread, there is less excuse for abuse or neglect through ignorance. Charities such as The Brooke, which seeks to educate people who are dependent on animals for their livelihood, still come up with some horrific stories: through custom and culture, the animals are often regarded as not being capable of feeling, their basic needs are overlooked, so-called cures barbaric. They are not respected, not *seen*. This of course means that the animals cannot give of their best, so the sordid cycle of deprivation on both levels, physical and emotional, continues. Breaking this by simple education about animal welfare produces spectacular results, a win-win situation whose repercussions are felt not only by the animals. For it's the same story: in kindness to them, we find kindness in ourselves.

* * *

Cruelty comes in many forms, from the unthinking to the wanton. The historical lack of empathy for animals described above is easily challenged and usually easily changed; once shown alternative measures to the hobbling, bleeding and overwork on poor rations that has been their custom, most people respond with enthusiasm, particularly when they realise that this prolongs the working life and capabilities of their animals. They begin to *see* their charges, with their individual needs and characters, rather than treating them as mere tools. It's particularly gratifying to find young children taking on caring roles, in the knowledge that this new understanding will be passed on down the line.

Deliberate cruelty is a different matter, and harder to fathom. It says much about the state of mind of the perpetrator, for you cannot abuse an animal without abusing yourself. But it is not straightforward. While living in Spain I was frequently appalled by the callousness of otherwise kindly people towards their animals, even those considered necessary to their livelihood. Dogs tethered on a short chain for their entire lives to bark at possible intruders on the flock or the home, horses ridden with bits that inhibited their free movement while inflicting horrible wounds, donkeys carrying loads that no carthorse could be expected to bear – these were common sights. And I learnt not to show my Spanish friends a particularly beautiful butterfly or a fascinating insect, for down would come the boot and *squish*, it was gone. These were not bad people, would not beat their children or abuse their wives, but empathy with animals was simply lacking in their cultural background.

Almost everyone who has dealings with animals on a regular basis will admit to moments of extreme irritation, lack of patience, things we would like to have done better – or not at all, as D.H. Lawrence records in his poem 'Snake'. Here, by a water trough in Sicily, Lawrence *sees* the snake, as he saw many other animals he so vividly describes, but his sight, his insight, is at war with his conditioning –

> The voice of my education said to me
> He must be killed.

And so, fuelled by fear, the tussle continues:

> Was it cowardice, that I dared not kill him?
> Was it perversity, that I longed to talk to him?

Finally he is overcome by a 'sort of horror' that leads him to hurl a log at it – and instantly regret doing so. Lawrence didn't kill the snake but he killed something in himself, and he ends, unforgettably,

> I have something to expiate;
> A pettiness.

Most of us will identify with this cry, and in most cases the pettiness – or worse – will have been triggered by fear. Some of the fears are deep-rooted in our unconscious, and snakes top the list of many people's phobias. To be fearful of sharks or crocodiles, even if you are unlikely ever to come across one, makes good sense: don't mix with them for you are likely to lose. Similarly, the howling of wolves, another primal fear-inducer, can be traced to our forebears who lacked the means to defend themselves. Less rational though even more common – three times more so in women than men – is arachnophobia, for most of our European spiders are innocuous and easily despatched if you can't stomach them; the fear must go back to the venomous varieties that scuttled in the dark corners of the caves of our ancestors. The dividing line between self-defence and wanton destruction is a fine one, and no one would blame a person for using force, or anything at his or her disposal, when in danger of serious attack from an animal, whether man-eating tiger or scorpion. But killing lightly, from unexamined fear, is a thoughtless act that demands you do *not see* what it is that you destroy.

* * *

There are more invidious forms of abuse. A dog that is regularly given chocolate and crisps and anything else from a crinkly packet, and who soon has difficulty walking, then breathing, is being abused. The cat 'kept safe' inside a house all day and night, never able to wander and explore and stalk, or stretch out full length in the sun; birds in cages, mice in cages, lions in cages, calves in cages – containment in almost any form is an abuse to the animal soul and can seldom be justified. But we do it and we justify it – and will go on doing so, and will seldom be labelled cruel.

To many people, the use of animals for scientific and medical research comes under the heading of abuse. It's a complex subject and one that arouses strong passions on both sides. Going to the root of it reveals much about our attitude towards animals, and it's not only animal rights protesters who feel uneasy about some of the arguments put forward in favour of animal testing. To argue that it is necessary in order to reduce suffering for humans and increase the quality of their lives does not hold moral water if it has the opposite effect on the animals involved.

It is a controversy that can't be resolved, and experimentation is likely to continue because it has produced advances in medicine that could not have been made in any other way. Therefore the best we can hope for is that the animals thus used are treated with care and kindness – that they are *seen* as sentient beings and not used as mere tools for our selfish ends, no matter what the suffering. There are signs of this happening, in terms of better living conditions and less invasive techniques in medical research; and as scientific studies become more sophisticated and it is possible to study individual cells rather than the whole animal, the need for dissection or distressing experimentation is reduced. The computer too now plays its part, enabling both the simulation and the sharing of information, thus avoiding unnecessary repetition of similar trials.

A recent procedure that can raise no objections among even the most squeamish is the training of dogs for medical detection. In Amsterdam a beagle called Cliff stalks hospital wards sniffing out patients infected with the superbug *Clostridium difficile*; he has an 80 per cent success rate. The In Situ Foundation, based in California, has trained dogs, predominantly German shepherds, to detect the odour of various types of cancer with spectacular success, and they are enlarging this to include other diseased states. They stress that it is not a replacement for conventional diagnostic techniques, but the dogs' ability to detect cancer in its very early state, or its recurrence, is an exciting discovery and one that looks set to develop. And to complete the feel-good factor, the animals they train are mostly rescue dogs, who are found normal homes as family pets but go to work – and enjoy it – during the day.

* * *

Squeamish or not, most people accept that our use of animals as meat is and will remain part of life for the majority of humans, though reducing the amount of meat we eat would be hugely beneficial all round. For both animals and people, butchering is a necessary sequel to hunting and husbandry. Most animals kill cleanly, if bloodily, and suffering for the prey is mostly confined to the fear of the chase – possibly alleviated by the adrenaline that boosts strength while dulling pain. Fortunately, the cat is rare in its desire to tease out the agony of dying. That's not to say that predators tearing at a carcass is a pretty sight, but it's an honest one – arguably more morally sound than an overweight Westerner pushing around his plate a piece of steak that he doesn't need or even want. And in the wild the remains won't go into the bin; there will be plenty of others waiting their turn to clean up after a kill, from lesser hunters to vultures to a myriad insects: all will be recycled.

'Butchery' is an emotive word for humans, with overtones lacking in the cleaner act of butchering. I experienced butchery when a Spanish neighbour offered to do the necessary with the sheep I had inherited with the farm. It was a horrific experience and not one to repeat, so when the next time came and two young Senegalese men who had recently arrived in the village offered to do it, I was nervous. But they were Muslims who had learned from their father how to slaughter according to Islamic law. They handled the animals kindly, prayed before killing them, and did it so quickly and efficiently that neither the sheep nor I had time to feel more than a brief moment of horror. This was not butchery, just a necessary job well done. And it is often stated, though less widely implemented, that those who choose to eat meat should once at least have watched a slaughter, so that we do not fall into the trap of thinking that meat comes plastic-wrapped and bloodless.

How about fish, which we are encouraged to eat because it is such healthy food, and which many people who are concerned about farmed animals are happy to eat? Fish don't come bloodless either, and the butchery is every bit as horrific, the wastage and concomitant damage far greater. Perhaps it is because we find it harder to relate to fishes, with their expressionless eyes and goofy mouths, living as they do in an alien watery world in which, however polluted, they are still able to move freely – until caught. For modern methods of fishery are devastatingly efficient and devastatingly random. Undesirable fish, along with many other species such as dolphins, turtles and even whales, seabirds, living corals and a multitude of other invertebrates, caught by the massive purse seines, bottom trawlers and longlines, are dismissively referred to

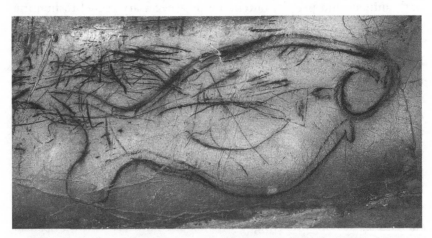

Fishes have been honoured, observed and doubtless eaten for thousands of years: this 1.5-metre long drawing of a flatfish, probably a halibut, was made around 18,000 BC and dominates other animals depicted in the Cueva de la Pileta, Malaga.

as 'bycatch'; the world's bycatch is estimated to be around one hundred million kilos every day, 40 per cent of the targeted fish catch. And the way these creatures die is as shocking as the statistics: hooked, entwined in netting, crushed, suffocated, decompressed (which means being turned inside out) – and then chucked back into a watery grave. Has this put you off eating fish yet? If not, try reading Jonathan Balcombe's *What a Fish Knows*. Balcombe, a lifelong lover of fishes (rather than fish – he makes the distinction) claims that not only are they sentient, they also have cognitive abilities. His case for this involves some mental acrobatics on his part, for fishes, even granted their individuality, appear to remain limited in their range of intellectual achievements. But that they are, indeed, individuals is important if we are to *see* them, rather than unceremoniously lumping them together as 'fish', a commodity to be killed no matter how, and eaten – with or without chips.

The loss of other sea creatures – the whales, sharks, turtles and dolphins 'mistakenly' scooped up and then dumped back into the water, in whatever condition – is having serious consequences for their numbers, and for the delicate balance of interdependent species. Nature allows for natural wastage, the over-abundance of one species providing for the survival of another, but not for wanton destruction on this scale. As Neaera H. ponders in *Turtle Diary*, the hundred eggs laid by a single sea turtle would be enough to ensure their survival from natural predators, 'but nothing ensures the turtles against the manufacturers of turtle soup. Three-hundred-pound turtles navigate the ocean and come ashore to be slaughtered for the five pounds of cartilage that gets sold to the soup-makers.' The fact that these creatures have swum 1,400 miles to get back to their breeding ground only compounds the injury. 'They're torn open and mutilated, left belly-up and dead or dying on the beach.' To commit such acts it must be necessary *not* to see the animal, in all its terrible beauty of broken body and purpose.

* * *

So much for abuse. But if the title of this chapter conjures up images of hefty Clydesdales straining at the plough, today these magnificent horses are more likely to be promoting beer at the Super Bowl. Animals sell. Often their use in advertising reflects the more dubious aspects of our relationship with them: sentimentality, anthropomorphism, romanticism

Corticelli Silk
WEARS LONGEST AND HOLDS STRONGEST

This 'pretty kitty' may be very different from the ruthless hunter of page 79, but it too can make a killing.

and exploitation, a sticky mix. Successful advertising plays into people's emotions, which is why any form of 'cuteness', animal or otherwise, appears to work. Even the Budweiser ads have recently been sugared up with the addition of that most apparently irresistible cuddly, the Labrador puppy (golden, of course), which sells not only beer but toilet rolls. Judging by current TV advertisements, appealing to the child in us invites us to loosen up, open our hearts and our wallets. Happiness, that elusive carefree emotion, is dangled before our eyes, along with the illusion that it can be bought: animals are used to buy into this illusion by taking us away from the daily grind into a land of so-called playful innocence (how far from the truth). Humour too lowers the defences; cartoon animals, often scarcely recognisable and far removed from their innate dignity, are commonly used in advertising, and animals with human characteristics are sure-fire favourites – think of the talking meerkats. Animals press buttons in us, and the promoters of advertising know how to use this to their advantage.

As do the charities. It's an old joke that people who fall out with their family members leave their money to cats' homes, but it seems there's some truth in it: animal charities are better supported in the UK than those for homeless, disabled or elderly people. Why is this? Again, sentimentality must come into it, and the charities use *that* to their advantage; after all, baby donkeys are hard to resist and maltreated ones harder still. But could it be something more? Are we trying to assuage a deep guilt we feel at the plight of animals that we have invited into our lives and then abandoned? Or is it that by the time people come to write their wills they are fed up with the jockeying of relatives and want to give to those who don't ask and are far removed from human wiles. If cuteness is used as exploitation, maybe that's righting the balance just a little on the side of the animals – and who could complain about that?

* * *

If the use of live animals, in the Western world at least, is decreasing as manmade products replace the skin, hair, wool and feathers that humans relied upon for so many millennia to keep themselves warm and dry, the value of these natural materials has seesawed. Shoes made of calfskin, sweaters of cashmere, duvets of duck down, all are more luxurious and now more sought after than their artificial equivalents. None of these is seriously controversial, but the use of other 'natural' sources is more difficult to justify. What if the shoes are made of crocodile skin, the soup of sharks' fins, the aphrodisiac of rhinoceros horn? Both the methods of killing, some of which are truly horrific, and the status of these fine animals makes their culling for such frivolous uses intolerable; this is truly abuse, not use, its only motivation greed and its continuation a stain on the relationship between humans and animals.

8
ENJOYING
ANIMALS

Given their freedom, sometimes too when in captivity, animals show a natural *joie de vivre* that is infectious. Birds in particular appear to be unburdened by worry, and watching even a scruffy town pigeon strutting his stuff on the pavement, let alone a lark ascending in full song or a swift scissoring the sky, triggers an involuntary lift of the heart for those who *see*. Their presence, physical and uncomplicated, shows up our human preoccupations for what they are, manmade and for the most part needless.

Enjoyment of animals is universal – and if you doubt that, think of the part they play in the early life of children. Even the most city-bound child, for whom fields and wide skies are an unknown land, will have a teddy bear on her pillow, will soon learn that C is for Cow and that they go *moo*, and will laugh at Donald Duck or shiver at the big bad wolf. As we have seen, writers of children's books have long exploited the power of animals over the young, and children all over the globe have grown up with animal stories and fables rooted in their own particular culture.

Children, very young children in particular, have a natural affinity with animals. They share much – vulnerability, spontaneity, lack of guile and a primitive language – and most children instinctively enjoy them, whether on paper, film, TV, the internet or, infinitely more affecting, the real thing. In the words of Toby already quoted, they are connected by an imaginary string from heart to heart, without the mind getting in the way. And most animals respond in kind, treating children gently and allowing them liberties of clumsiness or mishandling that they would never tolerate from an adult.

For children and young animals alike, play is central. It is how they learn, how they explore their world and their boundaries, how they grow up. Play is essentially fun – fun to do and fun to watch. Gambolling lambs, puppies chasing their tails and each other, kittens with a ball of string – all can be dismissed by the snooty as calendar kitsch, but in reality such antics are compulsive and few could watch without an involuntary smile, often a belly laugh as legs give way and everything tumbles in a heap. It is the spontaneity of animals enjoying themselves that sweeps away our usual measured responses, and within our enjoyment of them we may experience that rarer thing, joy.

But there is something more. If you begin really to watch these creatures, to *see* them in their play, you will soon realise that fun though it undoubtedly is, it is greater than just that. With play comes a sense of curiosity, a pushing of boundaries that is critical not only for the development of individuals but in the wider evolutionary process. I watch in fascination and some alarm as my dog Danny, needled by the onslaught of the puppy, takes her head in his mouth and gnaws at it, judging the strength of his jaws that could so easily crush it like an egg. Both – all three of us, in fact – are learning from this: finding how far to go, learning to trust each other. And if she finally oversteps the mark he will retaliate, not by crushing her skull but with a growl that clearly says, 'This time I mean it.'

Serious play – and that's not the contradiction it may sound – is most apparent in the more advanced species of mammals, those which are driven by evolution to explore and increase their capabilities. For these, scientists claim, play develops coordination, gives sensory stimulus, aids responses to stress; it stretches the young, puts them on their toes and stimulates them both physically and mentally. But that still doesn't *explain* play, which remains a mystery – arguably one of its most endearing qualities.

* * *

The enjoyment may be ours alone as the animals go about their business: the flash of a kingfisher on a darkening river, a mature cow turned girlish and kicking up her heels, a salamander revealed beneath an upturned stone – none of this demands anything of the animal but its beingness, but to the watcher it may bring a rush of pleasure that lifts a dull day and opens out the inwardness of preoccupation. Such sights are spontaneous, uncomplicated, and if looked for are seldom found. They are, quite simply, a gift.

Enjoyment of animals can be as close as the moggy on your lap or as far removed, in every sense, as the snow leopards in *Planet Earth II*. The latest cinematic techniques bring us images of staggering intimacy: movement, colours, textures, physical feats and ingenuity, in daylight and the depth of night, in jungle and back garden. Here aesthetics blend with scientific fact to show us what we have never been able to see before, though it is of a different nature to the intimate knowledge of those who have tracked, hunted, killed or kept animals from our very beginnings.

But for all their wonder, the animals we are now privileged to watch on screen lack one of the most central components of our pleasure in their company: touch.

It's easy to come over all soft when talking about petting anything, but a significant part of my enjoyment of the three dogs currently in my life is tactile. The daily ritual of exploring that face, the muzzle newly grey, the ears still velvet-soft, the underbelly, so trustingly exposed, soft in a different, damp sort of way. Lying on my back in the grass with them, which I do in most if not all weathers, is sensual too, their bodies pushed against mine, a paw laid over an arm, all of us at peace, sharing the most basic sort of being. (The world looks different, more picturesque, upside-down.)

It is not the mere presence of animals that soothes – though it does – but more specifically the physical act of petting and stroking, which through oxytocin releases feelings of love and empathy. People starved of physical love or lonely, children and old people in particular, are emotionally nourished when they pet dogs, cats, horses, donkeys – in fact anything warm, furry and alive that is prepared to submit to it. And one doesn't have to be needy, nor does the animal need to be furry, as demonstrated by a recent documentary about one of the oddest of all mammals, the pangolin. Looking like a giant pine cone, with claws that can break through concrete and the habit of curling up like a woodlouse when disturbed, these fully

Roxy, a pangolin taken in by Maria Diekmann, who devotes her life to defending these and other threatened species at her sanctuary REST, in Namibia.

scaled creatures are not everyone's idea of a cuddly companion. But here they are, being petted and hugged and carried about on people's shoulders or round their necks, with palpable pleasure on both sides.* This is the very essence of enjoyment: not ecstasy, not joy, but a deep contentment that feeds both the toucher and the touched, be it furry or scaly.

Some people – Brian Sewell, most notoriously – like to sleep with their dogs, and a recent survey found that 57 per cent of dog owners in the UK let their pets sleep on their beds or even under the covers. For me this is a step too far, and I prefer the pleasure we all get from greeting each other in the morning after a good night's sleep. (The same survey reported that 48 per cent said their sleep was interrupted at night as a result.) But dogs were used as bed-warmers by the Eskimos from the time of their earliest domestication, and Australian Aborigines still assess the temperature by the number of dogs needed to keep them warm at night: 'A very cold night might be a three- or even a four-dog night.' There is something deeply comforting about animal fur, even long after it has left the warm body that it came from. Perhaps this taps into our own once furry state when we groomed and petted and curled up with each other. Fake fur is not so inviting – nor does it smell the same, more proof of our animal nature.

* * *

Closely related to the sensual gratification we may experience with animals is aesthetic pleasure. Animal beauty is irrefutable, and we have seen what artists have made of it over the millennia. Beauty of form, of

---

* The pangolin is mercilessly hunted, now almost to the point of extinction, both as a delicacy and for its scales, which are used in Asian medicine.

movement, the unselfconscious physicality of animals as they luxuriate in being alive, now, at this moment – surely we must respond to this. Because animals when left to their own devices are comfortable in their own skins, this produces a physical ease, an elegance even, that for most people is synonymous with beauty.

But beauty is subjective, as is love, and many pets are beautiful to their owners however scruffy, lame and unappealing they may seem to others. I find my dogs beautiful in entirely different ways, even after their physical looks have waned and been replaced by a beauty of character and experience, but I don't expect other people to find them so (though I am ridiculously pleased when they do). Recently I sat by our lake in the darkening light, bats zinging round me and the odd trout rising, when there was a splash that was more than a trout and a head cut quickly through the water marking the point of a sharp V. Not the otter I desperately hoped for but a coypu, a semi-aquatic rodent almost as big as an otter but still very much a rodent. Why the disappointment, I wondered. Was it my lingering romantic image of Tarka, however dented by Charles Foster. Or prejudice against rodents in general, despite the charm of Ratty. Or was it to do with physical beauty, for the coypu is

The coypu, once sought for its fur, was introduced in the late nineteenth century to Asia, Europe and North America, where it is now widely regarded as a pest.

an ugly beast by anyone's reckoning. Uncomfortably pondering this, I decided to look for a different sort of beauty. How about the resilience of this coypu whose home was once South America, who has evolved, as I have, over millennia and is here now in a Breton lake, experiencing with me an evolutionary nanosecond in a habitat alien to us both; adapting and making a living and doing little harm except to the stems of aquatic plants, which in a small lake like this is actually a service. There is surely a beauty here, that of shared existence.

That 'beauty' is an unknown concept to the animals themselves adds extra weight to their attraction. There is a purity about animal looks that is absent from physical beauty in almost all human beings; it needs no mirror or cosmetics, and is without vanity. One might interpret the more exotic displays of some male bird, let's say a peacock, in his efforts to attract a mate as vanity, and claim that the lacklustre female is swept away by his colour and outrageous posturing, but he has to follow up with action or he will be dismissed, and to put any of this down to aesthetics on either side would be nonsense. It is Mother Nature who wields the paintbrush with wild abandon; the animals are merely her sitters, and we – if only we learn to see – the privileged connoisseurs.

* * *

Some readers may feel that I have exaggerated the enjoyment to be had from living with animals and glossed over the inconveniences and irritations. Of these there are plenty, so let's deal with them. The irritations occur on a daily basis, from minor (mud on the carpet), to medium (break-outs and break-ins), to major (a seemingly unresolvable behaviour problem). And inconvenience, yes. You are committed, tied – unless you can find and afford to pay other keepers – and responsible, both practically and emotionally, for these lives in your care. Like any other responsibility, it is one that can weigh heavily at times. Feeding and watering, shutting in and letting out, keeping warm and dry, is a relentless routine in which there is no place for lie-ins or excuse for bank holidays and hangovers. This is no light undertaking. On my very small 'farm', the following are regular demands and events: the sheep need to be moved from one field to another, occasionally caught for shearing or foot trimming – scope for much entertainment here; the duck shed must be cleaned out, and that's no fun; the geese fly over the fence and get chased, sometimes killed, by the dogs; the chickens cease laying, go missing, go broody or suddenly die for no good reason. Births and deaths are common, often violent. (One of the most upsetting was last year's relentless cull by crows of a batch of newly hatched Indian runner ducklings, eyes bright, feathers glowing, little legs pumping as they instinctively made their way from nest to pond.) But this is *life*. You can never get everything right, the best intentions are thwarted, it is muddy and bloody, and occasionally it all falls apart. So if you want your enjoyment of animals without any of that, you'd best stick to television documentaries.

Children get enjoyment from caring for animals, as well as useful lessons in life. Being responsible for a pet's welfare, learning that you can't simply forget to feed the goldfish, or put off – for too long – cleaning out the guinea pigs, or go away without knowing that someone will feed the cat; all these are a preparation for parenthood. In looking after, and at, animals most children will find solace, and not only of the cuddly sort. 'I suppose tropical fish were my religion,' wrote Julia Blackburn in a wry account of her childhood in a warring family. 'I performed something like prayer in front of the fish tank. I knelt there and stared into that silent world and slowly my racing heart would settle into a more steady rhythm and I would feel comforted.'

* * *

What of animal enjoyment? Anyone close to animals, whether domestic or wild, will vouch that they enjoy themselves. Whether in the boisterous games of youth or the sun-stretched leisure of old age, they know how to get the best out of life with the minimum of props. Lyall Watson gives a delightful account of the antics of a young peccary he rescued who learned to play games of catch with him, but also 'seemed perfectly happy to go away and play on his own, chasing his non-existent tail, whirling around in circles, trying to scratch his head with his hind foot, jumping up and down on the spot'. Not sophisticated stuff this, but clearly an animal enjoying being alive.

That most of the higher animals also enjoy each other is a given, just as they are at times irritated and bored by each other, like us. Mutual pleasure is likely to be physical, in playing, grooming, sometimes in intertwined sleep, and the enjoyment of my dogs as they greet each other is unmistakable even if it shows itself only in subtle body language, with

*The irrepressibly jaunty nature of piglets is perfectly captured
in this first-century bronze found in the ruins of Herculaneum.*

the merest touch of the nose or flick of a tail. Domestic animals, and some
wild ones, enjoy company, whether of their own kind or not. There's the
herd, of course, though this has probably more to do with the instinctive
need for self-protection in numbers than with actual enjoyment of each
other's company, and within the herd may lurk many rivalries.

Though most friendships are between animals of the same species,
strange affections occasionally spring up between species, and a scroll on
the internet will bring up a host of pictures of unlikely friendships, an
orgy of *oohs* and *aahs* as cats caress birds, birds play with dogs, dogs cavort
with cheetahs, tiger cubs curl up with chimpanzees. The eighteenth-
century naturalist Gilbert White describes how a cat who has lost her
kittens suckles an orphaned leveret, 'calling with little short inward notes
of complacency, such as [cats] use towards their kittens'. He goes on,
'Why so cruel and sanguinary a beast as a cat . . . should be affected with
any tenderness towards an animal which is its natural prey, is not so easy
to determine.' More usually it is the play-fighting common to young
animals of the same species that is uppermost, and that the animals
concerned enjoy these tussles is evident; they are spontaneous, even if
many are induced by proximity in artificial surroundings.

So animals enjoy each other, but do they enjoy humans? The answer seems to be yes, however surprisingly given how we've treated them over the years, given their lives today. Like us, the creatures are children of their time; they have known no other, except in their dreams and collective memories. And living as they do in the present, they adapt – as they must if their species are to survive.

What form their 'enjoyment' – somehow the word needs quotes – of us takes is hard to define, and obviously varies so widely that it may not be possible to defend. Maybe even domestic animals only enjoy what we provide, whether on the basic level of food and shelter or in the mistier realms of love. But there are some extraordinary examples of distances covered and difficulties overcome, of devotion that defies belief as animals (most often dogs) seek out or stand by their past keepers. Perhaps the best known of these is Greyfriars Bobby, a Skye terrier who followed his master's coffin to an Edinburgh churchyard and reputedly spent the next fourteen years on the grave. Although some have cast doubt on the story, it retains its status as the legend beyond all others of canine devotion and loyalty, and those qualities don't come without there

Faithful to the end. Edwin Landseer, The Old Shepherd's Chief Mourner, *1837.*

having been a great deal of enjoyment – undoubtedly mutual – in the relationship. Again, the internet will yield a wealth of similar vigils and of heroic journeys undertaken by dogs and, perhaps even more surprisingly given their independent nature, by cats. Nor can this be put down to a homing instinct, remarkable though that would be, since cats have made proven journeys of as much as 3,700 kilometres to find their keepers who have moved to another location. This is beyond human comprehension.

* * *

But humans too will go to extraordinary lengths to remain close to a favoured animal. A young Swede, Mikael Lindnord, was leading an adventure racing team in the Ecuadorian jungle when he became aware of a dog looking at him: 'I knew nothing about dogs. Never had one, never wanted one, but I could see that this dog was somehow special. It was as if he had some sort of inner calm, as if he knew stuff.' He shared some meatballs with the dog, then forgot about him as he prepared for what was to be the hardest part of the team's race. But the dog stayed with them, not just for the remaining two hundred kilometres of the gruelling course, but back to a life in Sweden with Mikael's family. There he remains today, a justified celebrity, a kingly dog named Arthur.

Rory Stewart came across a similarly regal dog while walking alone across Afghanistan. Described as a 'war dog' by the villagers who tolerated his presence only because he killed wolves, he was hungry, dirty, lame and without ears (cut off), teeth (knocked out) or tail – this was a dog with every reason to mistrust humans. 'I wondered how it felt never to have been stroked,' writes Stewart. 'His movements were somehow ponderous; there was no eagerness, no playfulness and no curiosity. I couldn't decide whether he was very depressed or very old or both. He looked over his shoulder and saw me. The stump of his tail moved slightly and he took a slow step towards me. I decided to take him back with me to Scotland.' Just like that! Named by him Babur after the Mughal emperor in whose steps Stewart was travelling, this heroic dog accompanied him for the next six hundred kilometres, at times having to be dragged, at others in the lead, attacked by other dogs, pelted with stones, both of them floundering through snow and often eating no more than dried bread for days on end. During all this, 'My relationship with Babur was developing. He was never a playful dog and he growled if I came near his food, but he was beginning to trust me.' The journey, as well as the trust, continued; but unlike Arthur's, this story does not have a happy ending. Babur died the day before he was due to board the plane that should have taken him to a well-earned life of ease in Scotland.

Dogs for sure, but few people could imagine becoming so attached to a chicken that they would choose it as travelling companion on a round-the-world sailing adventure. Yet that is exactly what a young Frenchman, Guirec Soudée, did when he set sail in 2014 with an ordinary-looking red hen called Monique. Although Soudée's given motive was that she would keep him supplied with eggs on his voyages – which she has – when they met a year earlier 'it was love at first sight', he says. And like any love affair, it has its moments: 'I won't lie, she can get on my nerves sometimes,' admits Soudée. But four years after embarking they are still together, and Monique has turned out to be not just any old hen, for she has learned to swim and surf in the Caribbean and to negotiate ice fields in Greenland, causing a stir as the first chicken ever to be seen there.

It may not be entirely coincidental that all three of the people mentioned here are exceptional, youthful adventurers with large hearts and open minds able to *see* what others might merely dismiss or pass by. Do these qualities draw animals to them? Is there an unspoken fellow-feeling between such courageous people and animals in their shared endurance of hardship out in the elements? Questions that cannot be answered definitively, but are worth pondering.

* * *

Animals enrich our lives in countless ways. As well as the utilitarian ones already touched on – providing meat, eggs, milk; leather, fur, feathers, wool; transport and pulling power; therapy and guidance – think what they have brought to religion and art, literature and language. It is hard to imagine the English language without words that we have taken either from animals or their attributes. Considering some of these I am struck by how many are used in a pejorative sense, often unjustly. You can be called a bitch, a cow or a rat, as well as being accused of wolfing down your food, having the manners of a pig, being catty or bitchy or flighty or bird-brained. You can be as timid as a mouse, as blind as a bat or as sly as a fox. It is well known that the Nazis deliberately dehumanised Jews by labelling them vermin, but less well recorded is that German prisonersof-war were likened by their Soviet captors to grey-green slugs, scorpions, plague-carrying rats and rabid dogs. Such words carry weight, and their meaning is unambiguous even to those unfamiliar with the animals concerned. Weasel words are there to mislead, chameleons change their allegiances, and watch out for that snake in the grass. On a more positive note one might be as busy as a bee or as brave as a lion, but these are the exceptions; most of our animal-related vocabulary equates the animal with the beast in us, and is none the less vivid for that.

As symbols and metaphors, animals even make their way into the alien world of politics and finance. Though both are now downplayed – the jokes and jibes presumably having gone stale – and they are no longer used as official logos, the donkey remains the emblem of the US Democratic party, the elephant that of the Republicans. And the bull and the bear continue to play their symbolic roles in stock markets around the globe.

* * *

Most people, even those who have little contact with or even liking for animals, dream about them. Here too they may represent what we like to call the 'animal' side of human nature: physicality, sexuality, aggression. In psychoanalysis, dream animals stand for the repressed subconscious, the instincts that we – or society – find unacceptable and we therefore attempt to bury. They indicate strong emotions, often those in need of attention, and as such act as messengers from the unconscious that are demanding to be listened to.

Certain animals turn up more often than others in dreams, and people may dream of types of animal – polar bears are a common example – that they know nothing about in everyday life. Shortly after writing this, I dreamed of a giraffe in the unlikely setting of our Breton garden; the small head on its impossibly long neck seemed to be smiling as it elegantly curled a long tongue around yellow fruits the size of melons, growing high in a dense evergreen tree. Where do such images come from? Well, since wild animals have surely dominated the dreams of our ancient forebears for hundreds of thousands of years in their roles as predator, prey and go-between for the gods, it is hardly surprising that they remain dominant in our subconscious, however lacking in modern everyday life. It seems that in dreaming, humans go back – or maybe go into – a part of their own nature that is closer to Nature, closer to their instincts and to their ancestors, with which animals provide a link. It may be arrogant to claim that this is a service the creatures provide, but it is true that just as they short-circuit our intellect in real life, so do they in our dreams. And for me, whose random daily encounters with animals so often bring an involuntary smile, those that appear in my dreams, equally unpredictable and elusive, leave a sniff of wildness and other-worldliness that permeates the day.

* * *

Our pleasure in animals creates a common bond, and many normally reserved people will start talking to each other at the vet's, at the zoo, while watching their dogs capering in the park. It's as if we *see* each other through our shared enjoyment, and inhibitions dwindle. (And unlikely things happen. I was once lent a flat in Paris after getting into conversation with a couple over their dog in a Hampshire pub; we never met again.) Animals animate; their uncontrived enjoyment, in life, in each other, is infectious, releasing our own. Which is why a life lived without any physical contact with animals seems to me a poor drab thing, a denial of our own physicality and potential for warmth.

But if you read this and feel deprived, living as you do ten storeys up in a city block, surrounded by concrete, neon and the noise of traffic, don't be downcast. Inspiration from animals can come in many forms, through reading, through those increasingly intimate documentaries, through the occasional visit to a wildlife park, or a holiday in a still wild part of the world. And by looking out for them even in an urban environment, noticing them as they slowly find their way in these new and often hostile surroundings, you will be enriched. The hawk on a high-rise, the fox raiding a dustbin, the butterfly trapped inside a window, all need as we do somewhere to rest, something to eat, a safe place to rear their young. These creatures are adapting to necessity and going about their lives unseen, until we begin to notice them. They are our fellow creatures, making their way like us through a life that is beset with difficulties but also full of wonder and delight if we can open our eyes to it – and to these animals that share our planet.

For their importance to us is paramount, and how we treat animals, our enjoyment and use of them, is an urgent matter that must not be dismissed as irrelevant in an increasingly technological world. In his visionary punt on the future of mankind, *Homo Deus*, Yuval Noah Harari writes, 'Some readers may wonder why animals receive so much attention in a book about the future. In my view, you cannot have a serious discussion about the nature and future of humankind without beginning with our fellow animals.' The creatures are part of our history, our culture, our world – an integral part of the greater pattern and order of life on earth, Nature, that has its own truths and is not corruptible by the need and greed of one species, however clever and manipulative that may be. So let us treat animals seriously, give them the respect and attention they deserve, and come to *see* them clearly and cleanly in their tragic vulnerability and inestimable value to us humans, as individuals and as a race. Our future, as much as theirs, hangs on this.

Notes

Introduction

Peter Matthiessen, *The Tree Where Man Was Born*, Picador, London, 1984, p. 86.   5

J.R. Ackerley, *My Dog Tulip*, New York Review Books, New York, 1999, p. 161; first published Secker and Warburg, London, 1956.   5

An abandoned wolf cub: Stef Penney, *The Tenderness of Wolves*, Quercus, London, 2007, p. 190.   6

Laurens van der Post, *A Far-Off Place*, Chatto and Windus, London, 1974, p. 159.   6

D.H. Lawrence, 'Elephants in the Circus', first published in *Birds, Beasts and Flowers*, Martin Secker, London, 1923.   6

John Berger, *Why Look at Animals?*, Penguin Books, London, 2009, p. 50.   6

1. Worshipping Animals

Laurens van der Post, *The Lost World of the Kalahari*, Penguin Books, London, 1958, p. 21.   9

This connection was confirmed: David Lewis-Williams, *Believing and Seeing: Symbolic Meanings in Southern San Rock Painting*, Academic Press, London, 1981, p. 78.   11

Van der Post, *Lost World*, p. 235.   11
Birds at Uitenhage (RARI LEE RSA Bek1), Rock Art Research Institute,   11
Witwatersrand University.
'the true author of every work': E.A. Wallis Budge, *The Gods of the*   13
Egyptians, vol. 1, Methuen, London, 1904, p. 414.
W.B. Yeats, 'Leda and the Swan', first published in *The Dial*, Boston MA, 14
1923.
James George Frazer, *The Golden Bough*, Macmillan, London, 1890, ch.   14
67.
Robert Storrie, 'Ambiguity in Northwest Coast Design', *From the Forest to* 15
the Sea: Emily Carr in British Columbia, edited by Sarah Milroy and Ian
Dejardin, Art Gallery of Ontario, Toronto; Dulwich Picture Gallery,
London, 2014, p. 105.
a recent report: *India Today*, New Delhi, 25 April 2017.   18
There's a story: Angela Fisher, *Africa Adorned*, Collins, London, 1984, 18
p. 15.
Proverbs 30: 24-28.   19
though the slaughter: Ezra 6: 17.   20

2. Depicting Animals

Kenneth Clark, *Animals and Men*, William Morrow, New York, 1977, p.   23
14.
crude, sticklike creatures: An exception to this are the Gwion Gwion   24
paintings of graceful, costumed human figures to be found in the Kimberley.
'They put paint': http://www.drakensberg-mountains.co.za/bushman-   26
rock-art.html.
Maikop bull, early Bronze Age (mid-3rd millennium BC), State Hermitage   27
Museum, St Petersburg.
'The Venus' and 'The Sorcerer', Chauvet-Pont-d'Arc Cave, Ardèche.   28
Pablo Picasso, *Minotauromachy*, 1935; *Blind Minotaur Led by Girl with* 29
Bouquet of Wild Flowers, 1934, Plate 94 of the Vollard Suite.
Piero di Cosimo, *A Satyr Mourning over a Nymph* (also known as *The*   30
Death of Procris), *c.* 1495, National Gallery, London.
White Hart, emblem of Richard II, back panel of the *Wilton Diptych*, *c.* 30
1395, National Gallery, London.
Diego Velázquez, *Las Meninas*, 1656, Prado, Madrid.   30
Henry Moore, *Sheep Sketchbook*, Thames and Hudson, London, 1980.   30
Henry Moore in conversation with David Mitchinson, *Henry Moore:*   30
Sculpture, Macmillan, London, 1981, p. 148.
William Hogarth, *The Painter and His Pug*, 1745, Tate, London.   31
Titian, *Man in Military Costume* (also known as *Giovanni dell' Acquaviva*), 31
c. 1552, Gemäldegalerie, Cassel.

3. DESCRIBING ANIMALS

Michael Bond, *A Bear Called Paddington*, Collins, London, 1958. 46

George Orwell, *Animal Farm*, Penguin Books, London, 1989, p. 90; first 47
published Secker and Warburg, London, 1945.

Herman Melville, *Moby-Dick*, Richard Bentley, London; Harper and 48
Brothers, New York, 1851.

Melville, letter to R.H. Dana Jr, 1 May 1850. 48

an 'ill-compounded mixture': Henry Chorley, *The Athenaeum*, London, 48
1851.

'a profound, poetic inquiry': Anonymous reviewer. 48

Ernest Hemingway, *The Old Man and the Sea*, Scribner, New York, 48
1952.

Fyodor Dostoyevsky, *Crime and Punishment*, first published as a series in 48
The Russian Messenger, Moscow, 1866.

Richard Bach, *Jonathan Livingston Seagull*, Macmillan, London, 49
1970.

Richard Adams, *Watership Down*, Rex Collings, London, 1972. 49

Richard Adams, Introduction to *Watership Down*, Scribner, New York, 49
2005, p. xvi.

T.S. Eliot, *Old Possum's Book of Practical Cats*, Faber & Faber, London, 50
1939.

Michael Morpurgo, *War Horse*, Kaye and Ward, London, 1982. 50

Michael Morpurgo, *The Scotsman*, Edinburgh, 6 January 2014. 50

as one critic said: Ben Brantley, 'A Boy and His Steed, Far From Humane 50
Society', *New York Times*, 14 April 2011.

4. Meeting Animals

artificial intelligence: https://futurism.com/animal-behavior-ai 51

Richard Mabey, *Nature Cure*, Pimlico, London, 2006, p. 22. 53

Hugh Lofting, *The Story of Doctor Dolittle*, Frederick A. Stokes, New York, 54
1920, ch. 2.

Michael Morpurgo, *New York Times*, 12 April 2011. 54

Russell Hoban, Introduction to *Household Tales* by Jacob Grimm and 54
Mervyn Peake, Macmillan, London, 1977.

'Magic Words' in Eskimo (Nalungiaq) translated by Edward Field, 55
Technicians of the Sacred, ed. Jerome Rothenberg, Sterling Lord Literistic,
New York, 1969.

scientific backing from several studies: See for example the work of the Swedish 55
researcher Kerstin Uvnäs Moberg and current research at Sussex University.

Jay Griffiths, *Tristimania*, Penguin Books, London, 2017, p. 60. 56

Christopher Smart, from 'Jubilate Agno', 1759-63, first published as 56
Rejoice in the Lamb: A Song from Bedlam, ed. W.F. Stead, Jonathan Cape,
London, 1939.

Vicki Hearne, *Adam's Task*, Skyhorse Publishing, New York, 2007, pp.   58
106-14 *passim*.

George Orwell, *Animal Farm*, Penguin Books, 1989, p. 31.   61

Lyall Watson, *The Whole Hog*, Profile Books, London, 2004.   61

There's research currently investigating: See for example Lori Morino   62
and Christina M. Colvin, 'Thinking Pigs: A Comparative View of
Cognition, Emotion, and Personality in *Sus domesticus*', *International
Journal of Comparative Psychology*, vol. 28, 2015; https://escholarship.
org/uc/item/8sx4s79c.

Temple Grandin and Catherine Johnson, *Animals in Translation*, Scribner,   64
New York, 2005, ch. 5.

Van der Post, *A Far-Off Place*, pp. 246-7.   66

5. Trusting Animals

Toby: https//www.dogsforgood.org.   73

Bill Thomas quoted in Atul Gawande, *Being Mortal*, Metropolitan Books,   73
New York, 2014, p. 122; taken from W. Thomas, *A Life Worth Living*,
VanderWyk and Burnham, St Louis MO, 1996.

Hearne, *Adam's Task*, 2007.   75

One trainer: Karine Vandenborre, https://horsefulnesstraining.com.   75

Alois Podhajsky, *My Horses, My Teachers*, Trafalgar Square Publishing,   76
North Pomfret VT, 1997.

If this sounds far-fetched: The benefits of humane treatment and   76
avoidance of stress for intensively farmed animals are now well
recognised, see for example P.H. Hemsworth, 'Human–animal
interactions in livestock production', *Applied Animal Behaviour Science*,
May 2003. Temple Grandin continues to work tirelessly to improve
conditions for animals reared on an industrial scale.

6. Hunting Animals

Paolo Uccello, *Hunt in the Forest*, c. 1460-70, Ashmolean Museum,   80
Oxford.

George Monbiot, *Feral: Rewilding the Land, Sea and Human Life*, Penguin   83
Books, London, 2014.

A.L. Kennedy, *On Bullfighting*, Yellow Jersey Press, 2000.   84

Carl Sandburg, 'Wilderness', *Cornhuskers*, Henry Holt, New York, 1918.   84

Ted Hughes, *Letters of Ted Hughes*, ed. Christopher Reid, Faber & Faber,   84
London, 2007, p. 658.

James Hillman, *Dream Animals*, Chronicle Books, San Francisco, 1997, p. 24. 85

Kes, directed by Ken Loach, Woodfall Film Production, 1969; based on   85
Barry Hines, *A Kestrel for a Knave*, Michael Joseph, London, 1968.

Helen Macdonald, *H is for Hawk*, Jonathan Cape, London, 2014.   85

T.H. White, *The Goshawk*, Jonathan Cape, 1951.   85

Isak Dinesen (Karen Blixen), *Shadows on the Grass*, Penguin Books,   87
London, 1985, p. 306.

John Fowles, *Wormholes*, Vintage, London, 1999, p. 301.   87

Foster, *Being a Beast*, pp. 146, 149, 155.   87

7. Using Animals

Brian Keenan, *An Evil Cradling*, Hutchinson, London, 1992, pp. 76, 287-   89
8.

Jean-Dominique Bauby, *The Diving-Bell and the Butterfly*, translated by   90
Jeremy Leggatt, Harper Perennial, London, 2008, pp. 104-5.

Desmond Morris, *Naked Ape*, Vintage, London, 1994.   91

www.thebrooke.org.   94

D.H. Lawrence, 'Snake', first published in *Birds, Beasts and Flowers*, Martin   95
Secker, London, 1923.

Jonathan Balcombe, *What a Fish Knows*, Oneworld Publications, London,   98
2016, pp. 219-21.

Russell Hoban, *Turtle Diary*, Picador, London, 1977, pp. 34-5.   99

8. Enjoying Animals

a recent documentary: BBC 2, Natural World, *Pangolins – The World's*   104
Most Wanted Animal, first aired 15 May 2018; see also https://www.
restnamibia.org.

Brian Sewell, *Sleeping with Dogs: A Peripheral Autobiography*, Quartet   105
Books, London, 2013.

Survey by Churchill Insurance, *Daily Telegraph*, 16 August 2012.   105

'A very cold night': Raimond Gaita, *The Philosopher's Dog*, Text Publishing,   105
Melbourne, 2002, p. 12.

Julia Blackburn, *My Animals and Other Family*, Jonathan Cape, London,   108
2007, pp. 24, 26.

Watson, *The Whole Hog*, p. 101.   108

Gilbert White, *The Natural History of Selborne*, ed. Richard Mabey,   109
Penguin Books, London, 1987, pp. 194-5.

Mikael Lindnord, *Arthur*, Two Roads, London, 2016, p. 86.   111

Rory Stewart, *The Places in Between*, Picador, London, 2014, pp. 153, 215.   111

Accounts of Guirec Soudée's remarkable voyages can be found on many   112
websites.

but less well recorded: Antony Beevor, *Berlin: The Downfall 1945*,   112
Penguin Books, London, 2003, p. 199.

Yuval Noah Harari, *Homo Deus*, Harvill Secker, London, 2016, p. 66.   115

Index

Numbers in *italic* refer to illustrations.